Insertional Mutagenesis Strategies in Cancer Genetics

Adam J. Dupuy · David A. Largaespada
Editors

Insertional Mutagenesis Strategies in Cancer Genetics

Springer

Editors
Adam J. Dupuy
Department of Anatomy & Cell Biology
Carver College of Medicine
University of Iowa
Iowa City, IA 52242, USA

David A. Largaespada, Ph.D
The Department of Genetics, Cell Biology
& Development
University of Minnesota
Minneapolis, MN 55455, USA

ISBN 978-1-4419-7655-0 e-ISBN 978-1-4419-7656-7
DOI 10.1007/978-1-4419-7656-7
Springer New York Dordrecht Heidelberg London

Library of Congress Control Number: 2010938734

Printed on acid-free paper

Springer is part of Springer Science+Business Media (www.springer.com)

Contents

Contributors

Keiko Akagi Mouse Cancer Genetics Program, National Cancer Institute, Center for Cancer Research, Frederick, MD 21702, USA; Comprehensive Cancer Center, Ohio State University, Columbus, OH 43210, USA, keiko.akagi@osumc.edu

Anton Berns Division of Molecular Genetics, Centre of Biomedical Genetics, and Cancer Genomics Centre, The Netherlands Cancer Institute, 1066 CX, Amsterdam, The Netherlands, a.berns@nki.nl

Katherine E. Berquam-Vrieze Department of Anatomy and Cell Biology, University of Iowa, Iowa City, IA 52241, USA, katherine-vrieze@uiowa.edu

Adam J. Dupuy Department of Anatomy and Cell Biology, Carver College of Medicine, University of Iowa, Iowa City, IA 52242, USA, adam-dupuy@uiowa.edu

Michal Dvořák Department of Molecular Virology, Institute of Molecular Genetics, Prague CZ–14220, Czech Republic, mdvorak@img.cas.cz

John Hilkens Divisions of Molecular Genetics, The Netherlands Cancer Institute, 1066 CX Amsterdam, The Netherlands, j.hilkens@nki.nl

Vít Karafiát Department of Molecular Virology, Institute of Molecular Genetics, Prague CZ–14220, Czech Republic, vit.karafiat@img.cas.cz

David A. Largaespada The Department of Genetics, Cell Biology and Development, The Department of Pediatrics, Masonic Cancer Center, The Center for Genome Engineering, The University of Minnesota, Minneapolis, MN 55455, USA, larga002@umn.edu

Michael D. Milsom Harvard Medical School, Children's Hospital Boston, Boston, MA 02115, USA, michael.milsom@childrens.harvard.edu

Lars U. Müller Harvard Medical School, Children's Hospital Boston, Boston, MA 02115, USA, lars.mueller@childrens.harvard.edu

Petr Pajer Department of Molecular Virology, Institute of Molecular Genetics, Prague CZ–14220, Czech Republic, petr.pajer@img.cas.cz

Vladimír Pečenka Department of Molecular Virology, Institute of Molecular Genetics, Prague CZ–14220, Czech Republic, vladimir.pecenka@img.cas.cz

Jesse D. Riordan Department of Anatomy and Cell Biology, University of Iowa, Iowa City, IA 52242, USA, jesse-riordan@uiowa.edu

Jean Roayaei Statistical Consulting Group, SAIC-Frederick, Frederick, MD 21701, USA, roayaei456@mail.nih.gov

Laura M. Rogers Department of Anatomy and Cell Biology, University of Iowa, Iowa City, IA 52242, USA, laura-rogers@uiowa.edu

Robert M. Stephens Advanced Technology Program, Advanced Biomedical Computing Center, SAIC-Frederick Inc., NCI-Frederick, Frederick, MD 21702, USA, stephensr@mail.nih.gov

Annabel Vendel-Zwaagstra Divisions of Molecular Genetics, The Netherlands Cancer Institute, 1066 CX Amsterdam, The Netherlands, annabelzwaagstra@yahoo.com

David A. Williams Harvard Medical School, Children's Hospital Boston, Boston, MA 02115, USA, dawilliams@childrens.harvard.edu

Ming Yi Advanced Technology Program, Advanced Biomedical Computing Center, SAIC-Frederick Inc., NCI-Frederick, Frederick, MD 21702, USA, yiming@mail.nih.gov

Chapter 1
Insertional Mutagenesis: A Powerful Tool in Cancer Research

Anton Berns

1.1 Introduction

Cancer arises largely as a result of mutations in genes regulating growth and differentiation of cells. In most tumors multiple genes are usually affected, resulting in either loss or gain-of-function of the encoded protein. The former category encompasses tumor suppressor genes whereas the latter belong to the oncogenes class. However, one should keep in mind that this designation can be context-dependent, and consequently a gene might act as an oncogene in one specific context and as a tumor suppressor in another. The number and nature of genes needed to drive cancerous growth can vary substantially and might lie between three and 10 and can greatly differ between different tumor types. If mutations occur in early progenitors, the cells might already be endowed with self-renewal capacity and therefore the additional mutations required to give rise to a malignant transformation might be very different from those needed to confer cancerous growth on further differentiated cells. The burning questions cancer researchers are facing are which combination of mutations are required to drive tumor growth in which of the cell types that constitute the tumor, and to what extent tumor growth is "addicted" to these mutations. Further insight into the mechanisms that promote metastasis and that can make tumor cells refractory to chemotherapy and targeted drugs are immediate next questions that we need to answer to develop more effective intervention strategies. This requires a detailed insight into how tumor cell proliferation and metastasis is controlled. Gene mutations critical for tumorigenesis have been found in a number of ways: by identifying transforming genes through DNA transfection experiments, by mapping recurrent translocations, amplifications, and deletions in chromosomal regions, by identifying the genes captured by acute transforming retroviruses, by defining the insertion sites of slow transforming retroviruses and transposons, and, more recently by the high throughput sequencing of cancer

A. Berns (✉)
Division of Molecular Genetics, Centre of Biomedical Genetics, and Cancer Genomics Centre, The Netherlands Cancer Institute, 1066 CX, Amsterdam, The Netherlands
e-mail: a.berns@nki.nl

A.J. Dupuy, D.A. Largaespada (eds.), *Insertional Mutagenesis Strategies in Cancer Genetics*, DOI 10.1007/978-1-4419-7656-7_1,

genomes. A number of general conclusions can be drawn from these studies. First, the number of recurrent mutations that confer a selective advantage to tumor cells is likely larger than previously thought. Second, a greater variety of genes than earlier suspected might have transforming potential when mutated or inappropriately expressed. Third, many of the genes contributing to tumorigenesis show a strong context dependence, and their expression levels might have to be within particular boundaries to effectively mediate the oncogenic effect, indicating that there is much more subtlety in the evolutionary process of tumor development than previously thought.

Insertional mutagenesis should be viewed as a "tool" in this context as it seems to play only a modest role in spontaneous tumorigenesis in man. However, it provides us with a potent methodology that can answer many of the burning questions. It can be used nearly in the same way as suppressor and enhancer screens previously performed in fruit flies and worms, to identify components of signaling pathways and to elucidate genetic interactions between pathways. This chapter is written to convey concepts rather than to provide the reader with experimental details. These can undoubtedly be found in many of the more specific chapters in this book.

1.2 Historical Perspective

Insertional mutagenesis has, in fact, a long history, although the underlying mechanism has remained obscure for more than half a century. It was at the basis of the high incidence of mammary tumors conveyed by cell-depleted milk of high-incidence mammary tumor strains as first reported by Bittner almost 70 years ago [4]. Insertional mutagenesis is also the cause of the high spontaneous lymphoma incidence recognized in some strains of mice in this same period ([30], #3801). The discovery of reverse transcription and oncogenic retroviruses focused much attention on the role of retroviruses in cancer. However, the difference between acute transforming retroviruses that carry oncogenes in their genomes and slow transforming viruses that can activate resident proto-oncogenes, was not immediately recognized. The first observation pointing in this direction was the observation that the spontaneous lymphoma development in the AKR strain was associated with an increase of the number of proviral copies in tumor DNA [3]. However, it took several years before the underlying mechanism, i.e., insertion of proviruses in the vicinity of proto-oncogenes or within tumor suppressor genes with their concomitant activation and inactivation, respectively, was elucidated.

In the first report to this effect it was shown that proviral insertions had occurred in the close proximity of an already well-known proto-oncogene—Myc in Avian Leukosis virus-induced tumors [16, 27]. While initially this provided information on the mechanism of action of these slow transforming viruses, it was quickly recognized that this could be used as a tool to identify new cancer-causing genes. The concept behind this is simple: Although the integration machinery of retroviruses and transposons does not cause fully random insertions—chromatin structure and sequence context undoubtedly leads to skewing—the process appeared sufficiently random to assume that insertion will occur in or near almost all genes when many millions of insertions take place in millions of cells. Occasional insertions near proto-oncogenes or within tumor suppressor genes might endow these cells with such a prominent selective growth advantage that this results in the outgrowth of clones that harbor insertions close to the same (putative) proto-oncogenes in independent transposon-induced tumors. The experimental strategy then is to induce tumors in a series of mice, e.g., by retroviral infection, and determine the insertion sites of the proviruses in the resulting, mostly clonal tumors. If proviral insertions or transposons cluster in the same region of DNA in tumors of independent mice more frequently than would be expected by chance, it is very likely that these insertions are associated with a selective advantage for the tumor cell. Such insertion clusters are called "Common Insertion Sites" (CIS). CISs are to be distinguished from "preferential insertion sites," the term that indicates that some loci might be occupied more frequently than others as a result of the interplay between chromatic structure, primary sequence and the integration machinery of the retrovirus or transposon. Preferential insertion sites constitute the background noise in insertional mutagenesis screens.

At the time the sequence of the mouse genome was not yet known and sequencing techniques archaic in comparison with current technologies, CISs had to be determined by cloning the flanking sequences of proviral insertions and using those as probes to ask whether changes could be found in the sizes of the corresponding genomic restriction fragments of DNA isolated from independent tumors using Southern blotting. Although laborious, this approach has identified many interesting new oncogenes that have provided important insights in the underlying mechanisms of tumorigenesis. The analysis of Mouse Mammary Tumor Virus (MMTV)-induced mammary carcinomas resulted in the discovery of the Wnt family of oncogenes critically important for stem cell self-renewal [28]. Another gene critical for stem cell renewal, Bmi1, was also identified by insertional mutagenesis [36]. A number of groups have exploited insertional mutagenesis to identify new oncogenes and tumor suppressor genes (for review, see [32]). In many instances it could be shown that genes carrying insertions within or nearby were deregulated or truncated by the insertions and had acquired oncogenic capacity or had lost tumor suppressor activity, thereby validating the robustness of the strategy. However, long-range effects have also been noted, making it sometimes cumbersome to identify the gene or genes near an insertion cluster conferring the selective advantage [20]. Altogether, the number of candidate oncogenes and tumor suppressor genes identified by insertional mutagenesis is impressive and in the same range as the number discovered by all other methods together, assuming the vast majority proofs to be genuine cancer genes. However, to identify these candidates required a relatively small effort in comparison with other approaches. Furthermore, it can serve as a valuable complementary strategy to the current deep sequencing initiatives of cancer genomes. Genes found by both methods are almost certainly involved in cancer and therefore insertional mutagenesis is very suitable as a cross-validation method for deep sequencing.

1.3 The Site of Insertion can Provide Mechanistic Insights

In a recent study in which we analyzed the insertion sites of around 500 Moloney MuLV-induced lymphomas, it appeared that a very substantial fraction of the 360 CIS that were found could be mapped to a position within the transcription unit. The remainder was found located mostly in the close vicinity, either up- or downstream of the supposed target gene. What type of conclusions can be drawn from the sites of insertion? The insertions up- or downstream of genes usually influence the level of the transcript, although translation can also be influenced, e.g., in case of promoter insertion in which transcription starts within the provirus resulting in an altered 5′ untranslated region that could influence translation efficiency. In many instances the insertions seem to act primarily as "enhancer" insertion supporting augmented transcription from a regularly used promoter (see Fig. 1.1). If multiple promoters are employed by genes—as is quite often the case—the enhancer insertion might selectively promote one transcript over the other thereby potentially also mediating a different biological effect. Insertions up- or downstream of genes therefore primarily change the expression level and rarely the nature of the encoded protein. The insertions found within transcription units fall in a number of different categories:

i. insertions in upstream introns and exons in a promoter insertion or enhancer orientation invariably give rise to an altered transcript but in most cases to the normal protein as most of these insertions do occur upstream of the predominant AUG initiation codon. Obviously, alteration of the 5′ untranslated region might change the efficiency of translation and can have a dramatic effect even if the level of the transcript is only modestly changed as has been observed for the activation of Lck [22].

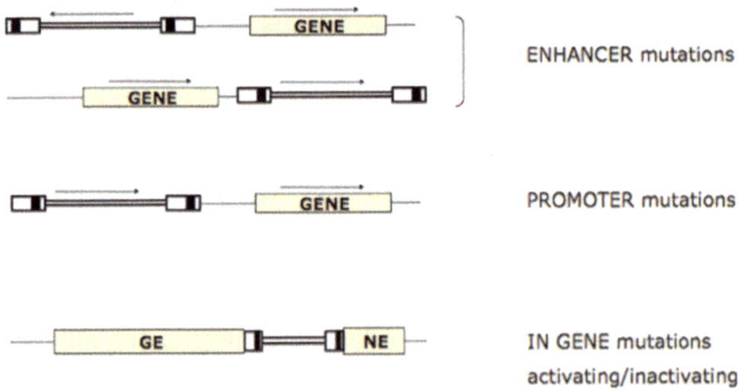

Fig. 1.1 Consequences of insertions on neighboring genes. A proviral insertion is provided as an example. In this case activations and inactivations of genes are possible through insertions that enhance transcription of the promoters normally driving transcription of the gene. Insertions in which the promoter of the provirus is used the 5′ sequence of the mRNA will be changed and this can lead to an altered translatability, or when inserted within the transcription unit to a different product that might either be more active (oncogenes) or has lost activity (tumor suppressors)

ii. Insertion in a noncoding region of a 3′ exon. Usually, the provirus can act here both as an enhancer and a transcriptional termination site and can so remove elements from the transcript that are targeted by micro-RNAs, thereby modulating either mRNA stability or its translation. This might explain the narrow domain in which proviral insertion site might be found. A good example is the CIS cluster near the translational stop codon in lymphomas with a provirally activated N-myc gene [34].
iii. Insertions in internal introns or exons in the same transcriptional orientation as the gene. This usually results in the expression of truncated proteins, either encoded by 5′ region of the gene or by the 3′ region of the gene starting translation either from an internal AUG or from sequences in the provirus with splicing to a downstream exon. Examples of both are found. Deletion of the carboxyterminal region of the Tpl2 (or Cot) protein kinase as a result of insertions within the last intron results in a constitutive active kinase since the carboxyterminal domain requires phosphorylation to permit activation of the kinase [5]. On the other hand, insertion in the Notch gene results in a mRNA that expresses the carboxyterminal region of Notch [17]. This is the business end of the protein and is normally produced through a gamma-secretase-dependent mechanism that relies on ligand-induced regulated intramembranous proteolysis. The carboxyterminal fragment that is produced then translocates to the nucleus where it becomes engaged in transcription regulation.
iv. Insertions within the transcription unit in either orientation can also result in a protein that acts in a dominant negative fashion or in its complete inactivation. This type of insertion is found in tumor suppressor genes leading to their inactivation. One might find insertions in both alleles of the tumor suppressor gene indicating that each insertion does confer substantial selective growth advantage to the cell.

Insertional mutagenesis is probably one of the most powerful methods to identify haplo-insufficient tumor suppressor genes as other genetic evidence for haploinsufficiency is often difficult to obtain. The different effects of proviruses or transposon on gene function is schematically depicted in Fig. 1.1.

1.4 Methodologies

Other chapters in this book will describe in detail specific new strategies in which DNA transposons are used that facilitate identification of the genes that are affected by the insertion. Therefore, I will limit myself describing a number of general features of insertional mutagenesis screens. A critical condition for successful screens is to generate a sufficient number of insertion events in order to have sufficient probability to achieve insertions near genes that can confer a selective advantage. The replication competent retroviruses are particular powerful in this respect as they enable multiple rounds of infection leading to the accumulation of proviral copies that can amount to often more than 20. This requires escape from interference by

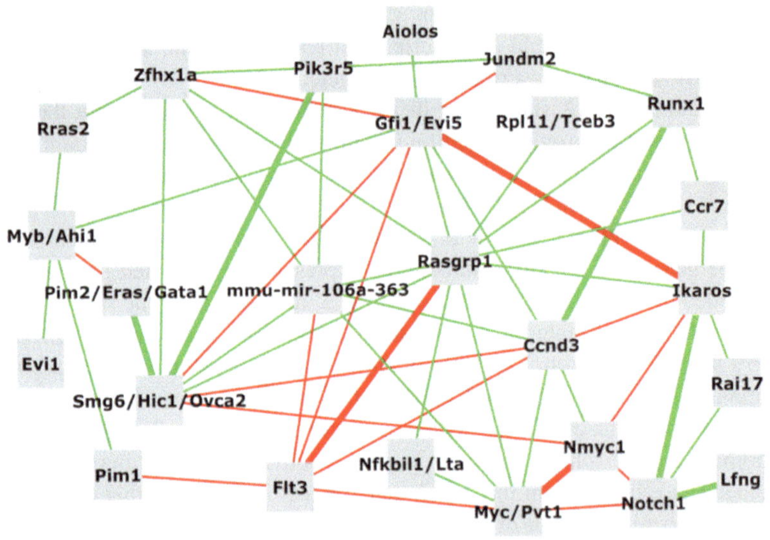

Fig. 1.2 Establishing interactions between common insertion sites. This diagram shows the interaction of a subset of the CIS found in the insertional mutagenesis screen described in Uren et al. [33]. CIS interaction network representing the co-occurrence or mutual exclusivity of the 20 most significant CISs. Co-occurring CISs are connected by *green lines* (*thin line*, $0.001 < p$-value < 0.05; *heavy line*, p-value < 0.001), mutual exclusive CISs are connected with *red lines* (*thin line*, $0.001 < p$-value < 0.05; *heavy line*, p-value < 0.001)

expression of envelope protein at the cell membrane of previously infected cells, and there are different mechanisms to reduce this interference, e.g., changing the envelope glycoprotein through recombination with endogenous viruses [29] and truncation of newly acquired proviruses due to mutations in the envelope gene. However, also replication defective viruses can give rise to a sufficient number of insertions to hit genes that can confer a selective advantage to the cell. Treatment of bone marrow of SCID patients with the therapeutic common gamma chain gene resulted in leukemias in a subset of the patients due to the insertion and concomitant activation of the LMO2 gene [15]. In the mouse, defective viruses have also been successfully used to mark oncogenes contributing to myeloid tumors [10]. In these instances the defective virus carried a gene that by itself can act as an oncogene. As a result, an insertion near a "collaborating" proto-oncogene would achieve two hits in one go. This might be one of the reasons why tumors ensue so effectively after infection with these oncogenic viruses. We can conclude from these observations that also with defective viruses a sufficient number of insertion events can be achieved to alter genes that can convey accelerated proliferation of cells and subsequent tumorigenesis.

When the number of infections is limited, it is likely important which subset of the cells are infected. Progenitor cells might be more prone to give rise to tumors as they might already have self-renewal capacity. This is particularly relevant when transposons are used that are inserted into the genome of all cells by germline modification and whose transposition is initiated by the cell-type specific activation of the transposase. The transposase might need to be activated in the right (progenitor) cell to permit tumor development. While this is a potential limitation, it also provides a unique opportunity to determine whether only particular cells in a tissue have the capacity to command tumorigenesis, e.g., by using distinct Cre strains with a highly specific target cell specificity. A high level of selectivity has also been observed with replication competent retroviruses that give rise to specific tumors, such as MMTV, Moloney murine leukemia virus, Friend murine leukemia virus and AKR endogenous ecotropic virus. However, one should realize that also the genetic background of the strains in which tumorigenesis is induced plays an important role both in the type of tumor that will arise and the latency period [12]. Although Moloney and Friend murine leukemia viruses are very similar in primary sequence, they give rise to different tumors in the same host due to small differences in their primary sequence [26]. These differences can influence many steps of the infection cycle of retroviruses and thereby influence the number of insertions in a particular cell type. With the DNA transposons that are now becoming the preferred insertional mutagen, we acquire more control over the process of insertional mutagenesis. This includes on the one hand the selective initiation of transposition in the cell type of choice at the desired time by Cre mediated activation of the transposase. In this setting the Cre is fused to a genetically modified estrogen-binding domain making it responsive to Tamoxifen, and driven from a cell-type specific promoter. Alternatively, Cre-mediated recombination can be used to make a transposase responsive to tetracycline regulation thereby providing the opportunity to switch the transposase on and off during tumor initiation and progression. This would potentially permit control over consecutive (in)activations of genes contributing to tumorigenesis and prevent undesired excision of transposons from loci that have contributed to the tumorigenic process. It is evident from the data that are now becoming available that insertional mutagenesis using DNA transposons works very efficiently and can give rise to multiple hits within a single tumor [11]. The added value of transposons is that they can be equipped with splice donor or splice acceptor sites making identification of the affected gene much less cumbersome. A drawback of the transposon systems is the preference to transpose locally in the vicinity of where the donor concatemer is inserted. First generation sleeping beauty transposons are known for this. However, the resulting skewing of insertion site can be simply ignored, and through the use of independent strains in which the donor concatemer is integrated in different chromosomes one can simply focus on the common insertion sites shared by the tumors in those strains.

One point important to note is that transposons, but even more so retroviruses, might elicit biological effects from the cell that are not related to insertional mutagenesis per se. In the case of retroviruses it is clear that additional mechanisms assist in the effective viral replication thereby facilitating insertional mutagenesis. The envelope might evoke a proliferative response [21] or, as is the case for MMTV, in which the synthesis of superantigens exploits the immune system to establish a chronic infection allowing infection of the mammary gland and transmission of the virus through the milk [1]. Similarly, transposition might elicit genetic instability or activate repair pathways requiring insertions in other gene sets in order to compensate for these "transposon-specific" effects. These "compensating mutations" might therefore mark genes that are unique for insertional mutagenesis model system without necessarily bearing relevance for tumors arising in man. Therefore, it is important to determine whether the "hits" of insertional mutagenesis screens play also a role in tumorigenesis in other settings, either experimentally or in human tumor cohorts.

1.5 Pairing Insertional Mutagenesis with Other Genetic Tools

Insertional mutagenesis can confer a selective advantage to cells. This advantage can be any feature one can select for. In such a way one can tune the system so that selection for particular features dominate, leading to the identification of genes whose altered expression or activity play a role in this process. The simplest tuning regards selecting for mutations in a distinct tissue or cell type. This can be achieved either by the nature of the infectious agent, e.g., MMTV gives rise to mammary tumors, MoMuLV to mostly lymphomas, and Friend MuLV to erythroleukemias, or by directing transposition to subsets of cells via manipulating the expression of transposases. In this way transposition can be activated in any cell lineage for which a tissue-specific promoter is available. Especially the DNA transposons that can be controlled by two levels of regulation (cell-type specific Cre making the transpose responsive to tetracycline) are particular powerful in this respect. In addition, transposons can be manipulated to skew their action pattern, e.g., to activate adjacent genes by equipping them with the appropriate enhancer elements in combination with a promoter and splice donor site, or to inactivate genes, by including a splice acceptor without an enhancer or promoter. This permits preferential identification of proto-oncogenes or (haplo-insufficient) tumor suppressor genes.

Insertional mutagenesis is also very suitable for use in sensitized screens. In this setting mice might already be predisposed to tumorigenesis and in that setting the prime question becomes which mutations are most effectively complementing the already present lesions. These lesions could have very different features, such as mutations in the Bloom's syndrome protein (Blm), permitting phenotype-driven recessive screens in diploid cells that made it possible to identify genes involved in mismatch repair [14]. We have used the strategy of sensitized screens by infecting tumor-prone Eμ-Myc or Eμ-Pim1 mice with MoMuLV and scored for new CISs that most effectively synergized with the predisposing lesions [35, 36]. One of the genes identified in this way was Bmi1, now known to be of critical importance for stem cell maintenance [31]. The same principle underlies the screen we recently performed in which we asked whether we would find skewing of insertion sites when comparing the CISs in lymphomas of wild-type, p19Arf–/–, and p53–/– mice.

Indeed we could find subsets of insertions that were highly enriched in each of these different genotypes [33] indicating unique interaction patterns with these resident lesions.

One can also utilize insertional mutagenesis to ask how to compensate for gene mutations that delay tumorigenesis. We have observed in the past that the co-expression of Myc and the protein kinase Pim1 results in a very strong synergism. Compound transgenic Eμ-Myc;Eμ-Pim1 mice are highly tumor prone, often succumbing already during embryogenesis from extensive lymphoid proliferation [37]. No other combination of oncogenes shows a similar high potency. This strong synergism suggested that Pim-controlled pathways are very important in tumors arising in Myc transgenic mice. We speculated that a screen in a Myc transgenic line lacking Pim might identify genes that can substitute for Pim. Therefore, we generated Eμ-Myc;Pim1–/–;Pim2–/– compound mutant mice and accelerated tumorigenesis by MoMuLV infection. Tumors that ensued showed distinct gene activations, among which Pim3, the last member of the Pim family, was found as a predominant target, as were other genes that likely can effectively compensate for Pim loss [24].

Using a different selection scheme one can also focus on genes involved in tumor progression or genes conferring drug resistance. The latter might be particularly relevant to understand the underlying mechanisms of drug resistance that often follows tumor regression imposed by cytotoxic or targeted drugs. A nice example has recently been published in which RUNX genes were identified as playing a critical role in conferring resistance to imatinib in chronic myeloid leukemia [23].

Finally, insertional mutagenesis does not necessarily have to act on genes directly but on any genetic element that can alter gene expression. In this respect it is noteworthy that expression of several micro-RNAs were found to be affected by insertions, in some cases with a very high incidence [19, 33]. Obviously, other not yet defined modulators of gene expression could be derailed by the insertional mutagenesis and, in fact, insertional mutagenesis might assist in identifying these.

Even without any specific predisposition one might simply ask the question whether parameters that differ among a group of animals in a particular experiment are associated with distinct insertion patterns. An obvious example is variation in tumor phenotype. In screens, some variation in tumor marker profiles might occur and be often associated with distinct CISs. But in fact many more correlations can be found, such as between sites of insertion and features like age of tumor onset, gender of the mice, or specific genotypic variations occurring in the experimental group.

I hope this convinces the reader that insertional mutagenesis is an extremely versatile and powerful genetic screening method that can be applied in an in-vivo setting. Especially now high throughput sequencing makes it possible to identify large numbers of insertion sites the approach will gain further momentum from the utilization of DNA transposons, which give rise to substantially more insertions per tumor cell clone than we are used to with replication competent retroviruses.

1.6 Mining "Hidden Information" in Mutagenesis Screens

Several studies either utilizing replication competent retroviruses or Sleeping Beauty transposons to accelerate tumorigenesis have shown that multiple CISs are often found in a single clonal tumor. This implies that insertional mutagenesis can catalyze tumor progression by mutating genes that effectively collaborate in the tumorigenic process. Since it is statistically nearly impossible to acquire almost simultaneously insertions near or in two or more cooperating genes, one has to assume consecutive insertions. The first insertion then should provide the cell with a selective advantage to promote its expansion, thereby generation a cell population in which a second collaborating "hit" becomes statistically feasible. The nature of the first "hit" has also consequences for the second "hit" that is selected: a second hit that complements the first mutation best has a higher probability to become enriched in the outgrowing clone. In the recent large-scale insertional mutagenesis screen we performed, this was precisely what we observed [33]. A number of facts are worth noting in this regard.

i. Particular combinations of insertions are found frequently. This is in line with the collaborating oncogene theory that has been proposed long ago [18]. Using a screen for lymphoma development, we did observe a high incidence of events leading to activation of components in the Myc, Ras and PI3kinase pathway. Most of the Ras-related mutations did not concern Ras proteins themselves—not surprisingly since insertional mutagenesis cannot induce point mutations and therefore the typical mutations found in Ras genes in many tumors cannot be achieved by an insertion in these screens—but rather other components stimulating the pathway such as the GDP/GTP guanine nucleotide exchange factor RasGRP1 that can indirectly enhance Ras-mediated signaling. Many other specific combinations were found. In fact, specific insertion clusters associated with the same proto-oncogene occurred often together with other distinct CIS. This might indicate that a particular level of expression is preferentially accompanied by insertions near defined other genes, likely because different levels of oncogenes expression require distinct collaborating events, e.g., in the case of Myc a number of CIS clusters can be found in its direct vicinity. Most if not all of these insertions, especially those that cluster close to the Myc gene, will likely result in enhanced Myc mRNA expression and Myc protein levels. Insertions at larger distance require closer scrutiny as they might influence other genes or control elements, such as CIS in the Pvt1 locus [2]. Nevertheless, the co-mutation spectrum even of CIS that evidently affect Myc is distinctly different for the various clusters.

There are several explanations for this observation:

i. The level of Myc expression is a determining factor in what are the most effective second lesions for oncogenes collaboration. The selection pressure for effective oncogene collaboration is in that model determined by the level of MYC protein.
ii. Alternatively, other oncogenic insertion might have preceded the insertion near Myc and determine what MYC levels are tolerated, e.g., without cells

undergoing apoptosis. One might envisage that insertions leading to impaired apoptosis would indirectly permit higher levels of MYC with concomitant higher proliferation and therefore positive selection for insertions causing higher MYC levels.

iii. Insertions in a particular region might be dictated by a different local chromatin structure resulting in a bias for insertions in particular subregions. Those biases might even be created by insertions near other genes that directly affect chromatin structure. Since a substantial fraction of the targets of insertional mutagenesis are in fact genes coding for chromatin-modifying proteins, this is an explanation that cannot be easily refuted.

Therefore, it is most likely that the specific combinations of collaborating insertions that are found in these mutagenesis screens reflect a well-tuned collaboration between these genes in tumorigenesis. An important question then is to what extent the tumor cell depends for its maintenance on this collaboration. If they do, drugs against either component, or both, might be particularly promising in treating tumors with co-mutations in the pathways in which these genes are involved.

iv. Some of the collaborations seem more straight-forward but nevertheless intriguing as they appear to enhance signaling in the same pathway. Examples include insertions resulting in overexpression of a normal Notch protein [33], which is frequently accompanied by insertions near Lunatic fringe, a glycosyltransferase known to modulate Notch signaling. The observation that insertions in the Notch gene directly giving rise to the active carboxyterminal region of Notch do not carry insertions near Lunatic fringe is in line with this explanation.

The occurrence of very specific combinations of insertions has also other practical ramifications: First, the occurrence of specific combinations reduces the chance that these insertions actually represent "preferential insertion sites" that are occupied due to the preferences of the integration machinery for chromatin or sequence context. Second, even if individual insertion might not occur frequently enough to reach statistical significance over background, specific co-occurrences almost invariable make these combinations highly significant. A good example represents insertions near the common gamma chain gene and Lmo2 (the combination also found in SCID patients that received retroviral gene therapy using a common gamma chain). A more obvious but highly specific combination represents independent insertions in each of the alleles of a tumor suppressor gene. This finding implies that insertion in one of the alleles has conferred already a selective advantage allowing selective expansion of cells carrying that insertion. The notion that p53 is a relatively rare target of insertional mutagenesis using MoMuLV might indicate that inactivation of one allele provides insufficient selective advantage, while this is clearly the case for disruption of both alleles as has been illustrated by the high lymphoma incidence in p53–/– mice. This would be in line with the observation that dominant negative mutations in p53 are very predominant. If this assumption were correct, one would expect p53 to be a more effective target for insertional inactivation in a P53+/– background.

Besides scoring for co-occurrences one might also check for lack of distinct co-occurrences as these could indicate that mutually-exclusive mutations fulfill a similar role in the tumorigenic process. This mutual exclusiveness is often seen for family members of oncogenes such as insertion near c-Myc and N-Myc are mutually exclusive, as are insertions near Pim1 and Pim2. While this argument is rather obvious for gene family members, mutual exclusiveness is also observed with genes that do not belong to the same gene family and this information can assist us in constructing wiring diagrams relevant for tumorigenesis.

1.7 Importance of Thorough Bioinformatics Analysis

The specific points made above with respect to extracting "hidden information" is only possible when using rigorous statistical analyses and advanced bioinformatics tools. The ability to identify multiple integration clusters in the vicinity of genes depends on methods that allow the investigator to vary parameters, such as the "window width"/"kernel width" and the statistical methods to reliably score for the co-occurrence in datasets with very large numbers of insertion sites [7, 8]. Since both the insertion machineries of the various retroviruses [9, 25] and DNA transposons [38, 39] differ and chromatin structure likely influences accessibility we encounter cold and hot DNA regions that might be poor or very well accessible to retroviral insertion or transposition. Ideally, one would like to establish the distribution of insertion sites in the absence of any selection pressure conveyed by the insertion-driven expansion of cell clones. This is complicated by two factors:

i. It is difficult to collect a large number of insertions with retroviruses in vivo as one has to retrieve cells shortly after infection and under those circumstances a relatively small fraction of the cells will carry retroviral insertions. However, time-controlled activation of transposase in transposon carrying animals does allow such analysis. It would be interesting to determine how random the initial insertions actually are and which regions in the genome register as "hot" and "cold".
ii. The insertion pattern is likely dependent on the differentiation stage of the target cell. This is for the moment an assumption and in vitro experiments using, e.g., ES cells that can be differentiated toward distinct lineages might give us a glimpse of how the variation in accessibility of chromosomal regions is dictated by the differentiation stage or other conditions of the cells.

None of the insertional mutagenesis screens performed up till now have included such background "controls" and this makes careful statistical analysis of the data even more important.

No doubt there is more information to be extracted from the datasets that we have so far. Factors that limit the power of these systems include uncertainty on whether co-occurrences observed in a tumor reflect co-occurrences within the same cell clone or in different clones composing the tumor. Especially when tumors develop very quickly there is the risk that they are oligoclonal. If a tumor used for analysis in fact harbored multiple cell clones this makes it more difficult to draw conclusions about co-occurrent or mutually exclusive insertion sites. This might be resolved by performing the analyses on single cells of a tumor or by choosing systems in which tumor development is relatively slow so that clonal tumors are more predominant.

Apart from improving the quality of the dataset there is also a lot to gain from more sophisticated analyses. Besides determining the interdependence of CISs, it is worthwhile to search for correlations between insertion sites and gene expression profiles using high density expression arrays. If those would show distinct correlations, one would like to determine whether similar expression profiles are also discernable in human tumors as this might point to the presence of corresponding driver mutations in human tumors.

1.8 The Added Value of Large Numbers

Insertional mutagenesis has been used since the early eighties and usually with small tumor sets (mostly less than 50). Recent larger studies, made possible by the availability of the complete mouse genome sequence and development of faster and cheaper sequence techniques, show that tumor panels of several hundreds tumors offer much more information, showing low frequency insertions and co-mutation frequencies that will be missed in small tumor panels.

The current methods in which insertions are exactly mapped to the nucleotide on the mouse genome sequence has a tremendous advantage over the "Southern blotting" approach of the early days of insertional mutagenesis screens. Different datasets were at the time almost impossible to compare unless identical digests and probes were used and even then one would easily miss significant CIS that were located just outside the diagnostic fragment used in the Southern blot analysis. Sequencing and exact mapping of the insertion sites overcome these limitations and allow pooling of datasets generated in different labs in different models. Now, the data have become cumulative, meaning that combined data sets provide more information and more power. This should serve as a stimulus to the field not only to produce large, complete, and thoroughly-curated datasets but also to put the information in a format that makes it easy to combine datasets produced in the different labs. In this respect it is very important to agree on a number of guidelines so that one can be assured that the datasets meet defined quality standards as to prevent loss of information when insertional mutagenesis datasets are combined.

1.9 Crossvalidation with Other Genetic Datasets

One of the elegant aspects of insertional mutagenesis screens is that it can identify at relatively low cost very large numbers of genes and control elements that likely play a role in cancer. At the same time one has to realize that a CIS does not unequivocally identify a gene or regulatory element as the culprit. This requires independent confirmation, e.g., by generating transgenic mice or by reproducing tumor proneness by overexpression or inhibition of the affected gene or controlling element in an appropriate cell type in vivo or in vitro using gene transfer. This has been used successfully in a number of cases for the validation of oncogenes acting in the hematopoietic system. Retroviral gene transfer into HSC with subsequent grafting in sublethally-irradiated mice then results in accelerated tumorigenesis. As a strategy to confirm the oncogenic potential conveyed by the large number of CIS identified in recent years, this appears a formidable task especially if one realizes that soon the number of putative oncogenes and tumor suppressor genes identified by insertional mutagenesis will probably amount to several thousands.

Interestingly, other methods to identify putative new lesions that play a role in tumorigenesis, such as the sequencing of cancer genomes, suffer from the same limitation. However, by directly comparing these different datasets, e.g., if mutations are found in genes that also are frequent target for insertional mutagenesis, in which both methods lead to either activation or inactivation of those genes, one can be fairly assured that the gene plays a role in tumorigenesis. In this way one can validate genes in both datasets and even gather additional information about their mechanism of action.

In comparing our recent dataset obtained in approximately 500 lymphomas and defining almost 350 oncogenes, tumor suppressor genes and control elements, we observed an almost 20% overlap with datasets of (putative) oncogenes/tumor suppressor genes generated by other methods in multiple human tumor types, including sequencing of cancer genomes. These numbers suggest that many oncogenic lesions have still to be identified in human tumors, although one should realize that not every genuine oncogene found by insertional mutagenesis might have a point-mutated, translocated or amplified counterpart in human tumors, simply because there might be other, more preferred routes to activate the pathway. The expectation is that with the application of transposons to induce a larger variety of tumors we will likely find much more overlap with the lesions found by these other methods. As sequencing of cancer genomes will soon become the method of choice to identify new lesions and knowing that the majority of the mutations found in tumor DNA actually represent background noise [13], the combination of high throughput insertional mutagenesis in different tumor models could be a very effective and economic way to unequivocally identify new cancer causing mutations.

1.10 Are New Cancer Causing Pathways to be Found?

In our recent large scale analysis of (500) lymphomas we asked the question whether among the close to (350) putative oncogenes and tumor suppressor genes evident candidates are found that on the basis of their sequence or known function would mark new signaling pathway not previously suspected to be involved in tumorigenesis. We analyzed the CIS for nearby genes that might contribute to tumorigenesis by completely different mechanisms. In view of the increasing evidence of an important role of changes in metabolism promoting of facilitating cancer development [6], we scrutinized the dataset for candidates that might act in one of its pathways. Although we cannot exclude that a small subset of the CIS actually targets new classes of putative oncogenes or tumor suppressor genes, we have not found clear evidence for this. Rather, the insertions seem to mostly influence genes that are known to act in the canonical cancer causing pathways. Time will tell whether some of the unknown genes act in canonical cancer causing pathways or point to new pathways whose role in tumorigenesis still has to be resolved. Possibly, transposon tagging leading to epithelial tumors might point to new pathways that were not recognized in studies focusing on lymphomas.

1.11 Conclusions and Recommendations

There is little doubt that insertional mutagenesis when conducted at a sufficiently large scale can provide a wealth of information on new putative oncogenes, tumor suppressor genes and other controlling elements present in genomes of higher organisms. No doubt, sophisticated bioinformatics analysis of the datasets can bring new important synergistic interactions between genes to light. The potency of the approach can be summarized as follows:

i. Insertional mutagenesis has the potential to identify very large numbers of new oncogenes, tumor suppressor genes including haploinsufficient tumor suppressor genes, and other control elements present in the genome of mammals.
ii. Scrutinizing the spectrum of co-mutations and mutually exclusive insertions, as well as the nature of mutations introduced within transcription units by CIS clusters, can provide important information about the underlying mechanism of tumor acceleration.
iii. Comparison of the "hits" with hit lists of other datasets, e.g., translocations, amplifications and deletions, and mutations identified by sequencing of cancer genomes will provide an "easy" but nevertheless robust route to validate components found in both datasets.
iv. In order to maximize the information from these datasets it will be critical to put in place accessible ways to share the datasets. This requires guidelines, both with regard to the quality of the datasets and the format in which the data are made available to the scientific community. Detailed information, not only about the statistical methods used to assign CIS status to a particular insertion cluster but also availability and accessibility to complementary information collected from those tumor samples, such as the marker profile of the tumors cells, and their expression profiles, will become of crucial importance to maximize the information.

Acknowledgements The author wants to thank Jaap Kool and Anthony Uren for their comments on the manuscript.

References

1. Acha-Orbea, H., Shakhov, A. N., & Finke, D. (2007). Immune response to MMTV infection. *Frontiers in Bioscience*, *12*, 1594–1609.
2. Beck-Engeser, G. B., Lum, A. M., Huppi, K., Caplen, N. J., Wang, B. B., & Wabl, M. (2008). Pvt1-encoded microRNAs in oncogenesis. *Retrovirology*, *5*, 4.
3. Berns, A., & Jaenisch, R. (1976). Increases of AKR-specific sequences in tumor tissues of leukemicAKR mice. *Proceedings of the National Academy of Sciences of the USA*, *73*, 2448–2452.
4. Bittner, J. J. (1939). Relation of nursing to the extra-chromosomal theory of breast cancer in mice. *The American Journal of Cancer*, *35*, 90–97.
5. Ceci, J. D., Patriotis, C. P., Tsatsanis, C., Makris, A. M., Kovatch, R., Swing, D. A., et al. (1997). Tpl-2 is an oncogenic kinase that is activated by carboxy-terminal truncation. *Genes and Development*, *11*, 688–700.
6. Christofk, H. R., Vander Heiden, M. G., Harris, M. H., Ramanathan, A., Gerszten, R. E., Wei, R., et al. (2008). The M2 splice isoform of pyruvate kinase is important for cancer metabolism and tumour growth. *Nature*, *452*, 230–233.
7. de Ridder, J., Kool, J., Uren, A., Bot, J., Wessels, L., & Reinders, M. (2007). Co-occurrence analysis of insertional mutagenesis data reveals cooperating oncogenes. *Bioinformatics (Oxford, England)*, *23*, i133–i141.
8. de Ridder, J., Uren, A., Kool, J., Reinders, M., & Wessels, L. (2006). Detecting statistically significant common insertion sites in retroviral insertional mutagenesis screens. *PLoS Computational Biology*, *2*, e166.
9. Derse, D., Crise, B., Li, Y., Princler, G., Lum, N., Stewart, C., et al. (2007). Human T-cell leukemia virus type 1 integration target sites in the human genome: Comparison with those of other retroviruses. *Journal of Virology*, *81*, 6731–6741.
10. Du, Y., Spence, S. E., Jenkins, N. A., & Copeland, N. G. (2005). Cooperating cancer-gene identification through oncogenic-retrovirus-induced insertional mutagenesis. *Blood*, *106*, 2498–2505.
11. Dupuy, A. J., Akagi, K., Largaespada, D. A., Copeland, N. G., & Jenkins, N. A. (2005). Mammalian mutagenesis using a highly mobile somatic Sleeping Beauty transposon system. *Nature*, *436*, 221–226.
12. Gilbert, D. J., Neumann, P. E., Taylor, B. A., Jenkins, N. A., & Copeland, N. G. (1993). Susceptibility of AKXD recombinant inbred mouse strains to lymphomas. *Journal of Virology*, *67*, 2083–2090.
13. Greenman, C., Stephens, P., Smith, R., Dalgliesh, G. L., Hunter, C., Bignell, G., et al. (2007). Patterns of somatic mutation in human cancer genomes. *Nature*, *446*, 153–158.
14. Guo, G., Wang, W., & Bradley, A. (2004). Mismatch repair genes identified using genetic screens in Blm-deficient embryonic stem cells. *Nature*, *429*, 891–895.
15. Hacein-Bey-Abina, S., Von Kalle, C., Schmidt, M., McCormack, M. P., Wulffraat, N., Leboulch, P., et al. (2003). LMO2-associated clonal T cell proliferation in two patients after gene therapy for SCID-X1. *Science*, *302*, 415–419.
16. Hayward, W. S., Neel, B. G., & Astrin, S. M. (1981). Activation of a cellular onc gene by promoter insertion in ALV-induced lymphoid leukosis. *Nature*, *290*, 475–480.
17. Hoemann, C. D., Beaulieu, N., Girard, L., Rebai, N., & Jolicoeur, P. (2000). Two distinct Notch1 mutant alleles are involved in the induction of T-cell leukemia in c-myc transgenic mice. *Molecular and Cellular Biology*, *20*, 3831–3842.
18. Land, H., Parada, L. F., & Weinberg, R. A. (1983). Tumorigenic conversion of primary embryo fibroblasts requires at least two cooperating oncogenes. *Nature*, *304*, 596–602.

19. Landais, S., Landry, S., Legault, P., & Rassart, E. (2007). Oncogenic potential of the miR-106-363 cluster and its implication in human T-cell leukemia. *Cancer Research, 67*, 5699–5707.
20. Lazo, P. A., Lee, J. S., & Tsichlis, P. N. (1990). Long-distance activation of the Myc protooncogene by provirus insertion in Mlvi-1 or Mlvi-4 in rat T-cell lymphomas. *Proceedings of the National Academy of Sciences of the USA, 87*, 170–173.
21. Li, J. P., & Baltimore, D. (1991). Mechanism of leukemogenesis induced by mink cell focus-forming murine leukemia viruses. *Journal of Virology, 65*, 2408–2414.
22. Marth, J. D., Overell, R. W., Meier, K. E., Krebs, E. G., & Perlmutter, R. M. (1988). Translational activation of the lck proto-oncogene. *Nature, 332*, 171–173.
23. Miething, C., Grundler, R., Mugler, C., Brero, S., Hoepfl, J., Geigl, J., et al. (2007). Retroviral insertional mutagenesis identifies RUNX genes involved in chronic myeloid leukemia disease persistence under imatinib treatment. *Proceedings of the National Academy of Sciences of the USA, 104*, 4594–4599.
24. Mikkers, H., Allen, J., Knipscheer, P., Romeijn, L., Hart, A., Vink, E., et al. (2002). High-throughput retroviral tagging to identify components of specific signaling pathways in cancer. *Nature Genetics, 32*, 153–159.
25. Mitchell, R. S., Beitzel, B. F., Schroder, A. R., Shinn, P., Chen, H., Berry, C. C., et al. (2004). Retroviral DNA integration: ASLV, HIV, and MLV show distinct target site preferences. *PLoS Biology, 2*, E234.
26. Mukhopadhyaya, R., Richardson, J., Nazarov, V., Corbin, A., Koller, R., Sitbon, M., et al. (1994). Different abilities of Friend murine leukemia virus (MuLV) and Moloney MuLV to induce promonocytic leukemia are due to determinants in both psi-gag-PR and env regions. *Journal of Virology, 68*, 5100–5107.
27. Neel, B. G., Hayward, W. S., Robinson, H. L., Fang, J., & Astrin, S. M. (1981). Avian leukosis virus-induced tumors have common proviral integration sites and synthesize discrete new RNAs: Oncogenesis by promoter insertion. *Cell, 23*, 323–334.
28. Nusse, R., & Varmus, H. E. (1982). Many tumors induced by the mouse mammary tumor virus contain a provirus integrated in the same region of the host genome. *Cell, 31*, 99–109.
29. Quint, W., Boelens, W., van Wezenbeek, P., Cuypers, H. T., Robanus Maandag, E., Selten, G., et al. (1984). Generation of AKR mink cell focus-forming viruses: A conserved single-copy xenotropic-like provirus provides recombinant long terminal repeat sequences. *Journal of Virology, 50*, 432–438.
30. Schwartz, S., Schoolman, H., & Szanto, P. (1956). Studies in leukemia. IV. The acceleration of the development of AKR lymphoma by means of cell-free filtrates. *Cancer Research, 16*, 559–564.
31. Sparmann, A., & van Lohuizen, M. (2006). Polycomb silencers control cell fate, development and cancer. *Nature Reviews Cancer, 6*, 846–856.
32. Uren, A. G., Kool, J., Berns, A., & van Lohuizen, M. (2005). Retroviral insertional mutagenesis: Apast, present and future. *Oncogene, 24*, 7656–7672.
33. Uren, A. G., Kool, J., Matentzoglu, K., de Ridder, J., Mattison, J., van Uitert, M., et al. (2008). Large-scale mutagenesis in p19(ARF)- and p53-deficient mice identifies cancer genes and their collaborative networks. *Cell, 133*, 727–741.
34. van Lohuizen, M., Breuer, M., & Berns, A. (1989a). N-myc is frequently activated by proviral insertion in MuLV-induced T cell lymphomas. *The EMBO Journal, 8*, 133–136.
35. van Lohuizen, M., Verbeek, S., Krimpenfort, P., Domen, J., Saris, C., Radaszkiewicz, T., et al. (1989b). Predisposition to lymphomagenesis in pim-1 transgenic mice: Cooperation with c-myc and N-myc in murine leukemia virus-induced tumors. *Cell, 56*, 673–682.
36. van Lohuizen, M., Verbeek, S., Scheijen, B., Wientjens, E., van der Gulden, H., & Berns, A. (1991). Identification of cooperating oncogenes in E mu-myc transgenic mice by provirus tagging [see comments]. *Cell, 65*, 737–752.

37. Verbeek, S., van Lohuizen, M., Van der Valk, M., Domen, J., Kraal, G., & Berns, A. (1991). Mice bearing the E mu-myc and E mu-pim-1 transgenes develop pre-B- cell leukemia prenatally. *Molecular and Cellular Biology*, *11*, 1176–1179.
38. Wilson, M. H., Coates, C. J., & George, A. L., Jr. (2007). PiggyBac transposon-mediated gene transfer in human cells. *Molecular Therapy*, *15*, 139–145.
39. Yant, S. R., Wu, X., Huang, Y., Garrison, B., Burgess, S. M., & Kay, M. A. (2005). High-resolution genome-wide mapping of transposon integration in mammals. *Molecular and Cellular Biology*, *25*, 2085–2094.

Chapter 2
Retroviral Insertional Mutagenesis in Mouse Models of Leukemia and Lymphoma

David A. Largaespada

Abstract Leukemia and lymphoma are cancers derived from cellular elements of the hematopoietic system. While they make up a minority of human cancer morbidity and mortality, the study of these cancers has illuminated many important aspects of cancer development and biology. In fact, the leukemias and lymphomas are among the best-studied and well understood types of cancer from a genetic perspective. In part, this may derive from the fact that these types of cancer are highly amenable to study using models in which mice are chronically infected with a retrovirus so as to induce or accelerate the disease. In this chapter, I have briefly reviewed the long and rich history of cancer studies using the murine leukemia viruses (MLV). Special attention has been paid to the replication competent MLV that typically cause cancer after a long latency and via insertional mutagenesis. This is followed by a discussion of the limitations of these models and suggestions for future work.

2.1 Introduction to the Murine Leukemia Viruses

The Murine Leukemia Viruses (MLV) are members of a large, naturally occurring group of type C gammaretroviruses that can infect rodents (reviewed in [22, 57]). The MLV were discovered many decades ago by careful observations, indicating a "filterable agent" could induce a malignancy of the lymphatic system in susceptible mice. These experiments were strong early examples that a virus could cause cancer. Like all retroviruses, the MLV carry an RNA genome that is reverse-transcribed into a double stranded DNA copy called a provirus. The integrated provirus is inherited by all daughter cells of the infected cell after cell division. The provirus serves as the template for the production of the messenger RNAs (mRNAs), which encode

D.A. Largaespada (✉)
The Department of Genetics, Cell Biology and Development, The Department of Pediatrics, Masonic Cancer Center, The Center for Genome Engineering, The University of Minnesota, Minneapolis, MN 55455, USA
e-mail: larga002@umn.edu

A.J. Dupuy, D.A. Largaespada (eds.), *Insertional Mutagenesis Strategies in Cancer Genetics*, DOI 10.1007/978-1-4419-7656-7_2,

the proteins required for virion production, and the viral RNA (vRNA) that will be packaged into new viral particles produced by the infected cell. The MLV are typically non-cytopathic viruses that do not result in lysis of virus producing cells. This fact, combined with the effects of integration of the provirus, described in more detail below, create the conditions that can lead to cancer development in infected animals.

The MLV can be divided into subgroups. The reader is referred to other comprehensive reviews on the structure, genetics, and distribution of the MLV, and other retroviruses [22]. Suffice it to say that early research showed MLV could be detected by their ability to induce syncytia formation in a cell line called XC, or focus formation in a cell line called SC [22]. Super-infection interference is a phenomenon that was observed when different isolates of MLV were used to infect the same cells in series. The second infection was blocked because of receptor occupancy by retroviral envelope protein produced by the first infection. This phenomenon, and the cell type tropism of the MLV, was used to show that these viruses could be placed into groups based on common usage of a cell surface receptor. The xenotropic MLV can infect non-rodent, but not rodent cells, despite the fact they exist in the genome of some mice. The ecotropic MLV can infect rodent cells, but not most other cell types including primate cells. The amphotropic MLV infect both rodent and primate cells. In most leukemia or lymphoma induction experiments done in laboratory mice, the ecotropic MLV are used. However, some of these studies have used amphotropic MLV, particularly in studies on acute myeloid leukemia (AML) induction.

All three types of MLV share a similar overall genome organization, but they have different envelope genes (env) that encode proteins, which use different cell surface receptors to gain entry into cells. The MLV are enveloped viruses, meaning they are surrounded by a lipid membrane bilayer derived from the infected cell that produced the virion and studded with env protein required to infect target cells. The env gene is expressed from a spliced mRNA, while the two other MLV genes—gag and pol—are expressed from the same full-length RNA that also serves as the genomic viral RNA (vRNA) destined to be packaged into new viral particles. Splicing between a splice donor (SD) and splice acceptor (SA) occurs with an appropriate frequency to generate env protein for virion production, while removing a special sequence called ψ, required to package the full-length vRNA into new virions. The gag gene encodes nucleocapsid proteins, while the pol gene encodes the reverse transcriptase and integrase enzymes required to produce the proviral DNA and integrate it into the genome of an infected host cell after viral infection.

The process of reverse transcription and integration of the proviral DNA is a complex process reviewed elsewhere [74]. However, the integrated provirus consists of the gag, pol, and env genes flanked by long terminal repeats (LTRs), which are sub-divided into the unique 3′ (U3), repeat (R) and unique 5′ (U5) regions. The U3 contains the enhancer and promoter sequences required to initiate transcription of the vRNA and spliced mRNA. The U5 region contains a polyadenylation signal. The U3 and U5 sequences, as well as the SD and SA sequences, are important in the process of insertional mutagenesis because they can alter the expression and processing of endogenous RNAs.

This review focuses on pressing issues related to the identification and study of leukemia and lymphoma genes identified via insertional mutagenesis by MLV. However, it is worth pointing out that replication-defective, partially deleted forms of the MLV, and other retroviruses, have been discovered that carry processed forms of endogenous proto-oncogenes. These viral oncogenes are found in acute transforming retroviruses (reviewed in [40, 47]). The study of these acute transforming retroviruses was pivotal in cancer research because they helped prove that cancer has a genetic origin and helped in the identification of many important targets of mutation in human cancer, such as the *RAS* oncogenes and *MYC*. The acute transforming retroviruses can be distinguished from the replication competent, slow transforming retroviruses, including many MLV, that are the subject of this review.

2.2 Viral and Host Determinants of Disease

Much prior research on the MLV was done to discover and map viral and host determinants of disease induction by the MLV. These studies are less often performed now, but were very critical from a practical standpoint because they taught lessons about the mechanisms by which the MLV cause cancer in the first place. It is worth pointing out that studies have allowed scientists to create new chimeric MLV useful for studying particular types of leukemia (for example [77]).

The murine host strain can influence disease by restricting infection or replication of the MLV. The best studied example is the *Fv1* gene, which exists in laboratory strains of mice in at least two different forms: *Fv1*^{b}, which restricts infection by b-ecotropic MLV; and *Fv1*^{n}, which restricts infection by n-tropic MLV [3]. Thus, most laboratory strains of mice carry one or the other allele and can be infected by only one or the other type of ecotropic MLV. Beyond the effects of *Fv1* and other restriction factors, the host strain background interacts with the strain of MLV to dictate the ultimate disease course. Thus, the same MLV can induce different diseases in the different inbred strains of mice. These differences could reflect inherent differences in susceptibility to various forms of leukemia in different inbred strains of mice. For example, a typical b-ecotropic MLV causes acute myeloid leukemia in BXH-2 strain mice [3, 25]. The susceptibility of BXH-2 strain to AML induction in this model, rather than other forms of leukemia, has been tentatively traced to a unique BXH-2-specific germline mutation in the *Icsbp1* gene in this strain [62].

One often ignored host strain determinant of leukemia induction by MLV is the development of other replication-competent viruses in the infected host via recombination. It is common for mice that are chronically infected with an ecotropic MLV to develop recombined MLV called mink cell focus forming (MCF) viruses [57]. The MCF are generated by recombination events between ecotropic MLV and vRNAs produced by endogenous MLV which lead to replacement of env gene and LTR sequences. The MCF utilize a different cell surface receptor for infection than do ecotropic MLV and can escape superinfection interference by ecotropic MLV. The MCF may drive leukemogenesis by allowing more proviral insertion mutations to accumulate in pre-neoplastic clones. The details of the strain-specific effects of some MLV are in general not known and useful models of MLV-driven leukemia are generally found, or tested empirically, rather than being deliberately created.

Viral determinants of disease specificity have also been discovered. Often this has been done by switching sequences and genes within one MLV, with those of another MLV to determine what sequences determine disease specificity. The determinants of leukemia specificity have often been shown to map to the LTRs [51] and within an LTR, to the enhancer sequences in the U3 region [77]. These studies have suggested that the ability of some MLV to specifically induce one type of leukemia, and not another, may be due to the activity of the LTR. If an LTR promoter were especially active in one cell type versus another, it is possible that viral spread within that cell type would be greater, and that activating mutations would be more likely. The MLV LTR is a major driver of gene activation in leukemia induction studies because it can enhance transcription from endogenous promoters and drive chimeric RNAs composed of viral and endogenous mRNA sequences. These events are common in the activation of many endogenous genes by MLV insertion. It should be noted that some mechanisms of MLV mutagenesis do not seem to require LTR activity. In some cases, MLV insertions seems to alter the splicing and processing of an endogenous mRNA encoded by genes that have suffered proviral insertions. These events can decrease the expression of a tumor suppressor gene [35], or stabilize an oncogene mRNA by replacing 3′ UTR sequences that contains destabilizing motifs [67, 69]. The viral determinants of MLV leukemia induction are in general not well enough known to predict *a priori* what sequence alterations to the viral sequence should be made to produce specific types of leukemia, let alone specific types of cancer in mice. Since MLV, and other gamma retroviruses, can only efficiently infect cells undergoing mitosis [43], there must be inherent limitations in the tissue tropism that can be built into the MLV. Nevertheless, it would be highly desirable to generate new recombinant MLV with useful properties, such as the ability to induce each of the forms of leukemia and lymphoma that characterize human disease. For example, among the eight or more French-American-British (FAB) subtypes of AML known to occur in humans, not all can be induced in mice. It would be useful to have MLV-induced models of M6 and M7 AML, but the alterations that might allow this are not easy to predict. Despite these issues, there have been engineered MLV that can induce a spectrum of leukemia not usually seen with the parental viruses and which have other useful properties.

The MOL4070LTR virus is a chimeric MLV containing U3 LTR sequences from the amphotropic MLV called 4070A and other sequences from the Moloney-MLV [77]. The 4070A virus has been shown to induce AML in some strains of mice [76]. However, this virus is restricted in the number of common laboratory strains in which it can be used, as 4070A is an n-tropic MLV. Moloney-MLV is an nb-tropic MLV and so can infect and replicate in mice that carry either the $Fv1^n$ or $Fv1^b$ alleles common to various laboratory strains [77]. However, Moloney-MLV induces almost exclusively T cell leukemias in mice [20]. The MOL4070LTR virus has the nb-tropism of Moloney MLV, which is determined in the gag gene sequences [54], and the ability to induce AML in some strains of mice, from the 4070A parent virus, a trait apparently largely influenced by the U3 sequences [77]. This virus has been shown to induce AML and ALL in various laboratory strains of mice, such as FVB/n and 129/SvJ, and in various leukemia-susceptible transgenic mice [6]. The MOL4070LTR virus thus seems to be generally useful for finding cooperating mutations in AML and ALL using strains commonly used to make transgenic mice. For example, this virus was used to find gene mutations that could cooperate with expression of the *NUP98-HOXD13* oncogene [59]. We have used the MOL4070LTR virus to find gene mutations that could cooperate with expression of the *MLL-AF9* oncogene in AML and ALL development (unpublished observations).

Others have replaced the enhancers within an MLV to alter its disease specificity. The PyF101 + Mo M-MuLV consists of a Moloney-MLV in which the viral enhancers in the U3 region are replaced with those from the SV40 early promoter [21]. This virus induces myeloid and T lymphoid leukemia in mice. These kinds of experiments are perhaps hampered by our inability to fully predict what kind of disease will result, and from the fact that the alterations that are made could impair the ability of the virus to replicate. Despite these concerns, the example provided by the PyF101 + Mo M-MuLV suggest that perhaps hematopoietic lineage-specific enhancers could be included in the U3 region of a recombinant MLV to achieve specific leukemia phenotypes in mice. Transposable elements, such as *Sleeping Beauty*, provide an alternative to this approach [60]. These are described in a different chapter of this book.

2.3 Challenges in MLV Mutagenesis Studies

There are many challenges in MLV mutagenesis studies. Among the first to consider, when starting a project, is the number of independent leukemia samples that should be generated for study. As has been observed before, “quantity has a quality all its own.” Unfortunately, experience has shown that more statistically significant CIS and associated genes are discovered as more leukemias are studied, even when one approaches and surpasses 100 individual leukemias. The large number of leukemias required for identification of rare CIS—involved in <10% of cases, can be cost prohibitive to generate. Our usual goal is to obtain 60+ mice of each experimental group expected to get leukemia. This number is based on having the precision to detect any common insertion site present in 10% of the tumors balanced against the cost and expense of generating and aging these cohorts. If we assume that each mouse will develop a malignancy with on average one cloned insertion per tumor the R statistical package (http://www.r-project.org) calculates the binomial probability of missing a common insertion site, with a 10% true insertion frequency, as 1.38% at this sample size. This is a conservative estimate, as usually many more than one insertion per leukemia is recovered. As we have seen, not only are many mice needed, but slow transforming retroviruses typically induce leukemia after 5 or more months. It can often cost $20-30,000 just in mouse animal housing charges for a project like this one—assuming it lasts 1 year.

One other complication of studies of this sort has to do with recovery of proviral insertion sites. Many different approaches to accomplish this have been described in the literature in the past. Methods based on the use of primers specific to the MLV and which use the polymerase chain reaction (PCR) are currently preferred [37]. This issue is complicated by the fact that any MLV induced leukemia is actually a collection of some number of related, but genetically distinct clones. Some insertion mutations are present in only a subset of leukemia cells. The significance of these sub-clonal insertion mutations is not entirely clear. Work by investigators, such as Dr. Philip Tsichlis, clearly shows that new subclones with new growth-promoting proviral insertion mutations can be identified in MLV induced leukemias as they are passaged in vivo or in vitro [55]. Recently, ligation-mediated PCR (LM-PCR) based methods for proviral or transposon insertion site recovery have been linked to "single molecule" or "pyrosequencing" approaches, such as that made possible by the Roche GS-FLX machine [60, 64]. Thus, hundreds of thousands of amplified provirus/cellular genomic DNA junction PCR products can be sequenced, from 100 or more leukemias, en masse, without the need for cloning them beforehand. By using bar-coded PCR primers for the secondary PCR in the LM-PCR reaction, it is possible to determine, after sequence analysis, which insertions were derived from which individual leukemia. A benefit of this approach is that the number of times any one specific insertion sequence is read may be related to the clonal abundance of that insertion mutation in the leukemia. So, two sorts of data are obtained: the identity of all insertions and their relative abundance in the tumor clone. However, for these data to be most meaningful, it is most desirable for the recovery of insertion sites to be saturating. Therefore, we recommend that insertions are amplified and sequenced from both ends of the integrated provirus and that multiple restriction enzymes be used to generate the junction fragments that will be amplified. We refer the reader to several methods articles for details on this kind of approach [34, 64, 80]. A few complicating issues bear mentioning, however. It is important to keep in mind that when amplifying and sequencing proviral insertion sites the fact that they are flanked by LTRs presents a special problem. One must introduce a restriction enzyme digestion step to avoid recovery of internal proviral sequences downstream of the 5′ LTR, when amplifying "right-hand" junction fragments downstream of the 3′LTR. Similarly, a restriction enzyme digestion step is introduced to avoid recovery of internal proviral sequences upstream of the 3′ LTR, when amplifying "left-hand" junction fragments upstream of the 5′LTR. We have developed methods for proviral insertion site amplification using LM-PCR that rely on the use of blunt-ended restriction enzymes, or sticky-ended restriction enzyme digestion followed by filling in the overhangs with *Taq* polymerase, thus generating products with a single adenosine (A) overhang [80]. In each case, after generation of junction fragments, they are then ligated to linkers and amplified using two rounds of PCR. In nearly all of the techniques used to generate amplified LM-PCR products the linkers are double-stranded and some provision is made to prevent PCR amplification of "linker to linker" fragments and allow "linker to proviral LTR" PCR amplification. This is accomplished by using as a linker primer, a sequence that has no homologous sequence to anneal to until and unless the LTR-specific primer has annealed to a template junction fragment allowing the PCR polymerase to generate a product that extends into the linker and generate sequence complementary to the linker-specific primer. This can be done using two primers of unequal length to create a partially double-stranded linker, in which the 3′ end of the shorter primer is blocked to prevent the polymerase from using it to create a fully double-stranded primer during the PCR reaction, which would cause the generation of many contaminating "linker to linker" PCR products.

It is possible that if one shears leukemia genomic DNA, "blunt ends" all the fragments using a polymerase, and then does LM-PCR using blunt-ended primers, it may be possible to recover more insertion sites than using restriction enzymes to generate fragments. Innovations such as this one will allow greater recovery of all MLV proviral insertions that are present. In the future, it seems possible that MLV mutagenesis studies would benefit from whole genome re-sequencing, once sequencing technology becomes cheap enough to perform on a large number of samples. Such data may reveal cooperating mutations not induced by MLV insertions, but by other mechanisms such as translocations, point mutations or deletions. In any case, very robust bioinformatics analysis of proviral insertion site sequencing results is required before determining what significant, recurrently occurring, insertion site mutations have been discovered. It is important to remember that LM-PCR may generate products from endogenous retroviruses. In some cases, these are well known retroviruses as in the endogenous b-tropic MLV in BXH-2 strain mice [80]. In other cases, they can be recognized because identical insertion sites are amplified from multiple independent mice. In addition, the sequence analysis algorithm must be designed to recognize bona fide MLV insertion sites by recognition of LTR sequences downstream of the second LTR-specific primer, followed by genomic sequence mapping with high confidence to one unique position in the mouse genome. Once a non-redundant set of insertions is defined, the definition of common sites of insertion (CIS) follows.

Several methods for determining which genomic loci have suffered a statistically significant number of independent insertion mutations have been published [13, 78]. However, no clear consensus exists for the correct methods. It was assumed in the past that MLV integration was a random process, not influenced by the genomic architecture. This meant that Monte Carlo simulations could be performed using a number of randomly selected genomic positions and calculating the expected fraction of random or background CIS detected in different window sizes and these were used to select cut-offs for significant CIS observed in actual data [39]. Other approaches calculated CIS based on the assumption of a Poisson distribution of insertion mutations in the genome [42]. However, more recent analyses show that MLV insertion is not random and is heavily influenced by the presence of genes, with insertions near the start site of gene transcription being preferred. Thus, a null set of MLV insertions has been obtained which provides a different estimate of statistically significant CIS [78]. It has been estimated that up to 2/3rds of the CIS defined in the Retroviral Tagged Cancer Genome Database [1] can be explained by the null hypothesis, that is, the lack of any selective pressure for their mutation by MLV insertion [78]. It was recommended that CIS defined by four or fewer insertion mutations be considered cautiously [78]. The problem of background false positive CIS is compounded by the large data sets obtained using LM-PCR and high throughput sequencing—resulting in definitions of CIS requiring large numbers of insertions within small regions of the genome. Thus, CIS in tumor suppressor genes that are very large and where insertions in various parts of the gene could disrupt them might be missed. Similarly, if MLV can effect the transcription of a gene at a long distance, CIS might be missed in a large dataset of insertions because they are spread out over a large area. To correct for these problems, Dr. Wessels and colleagues in the Netherlands have defined CIS using an entirely different and sophisticated method [13]. In this method, an expected distribution, called a kernel density estimate, of insertion mutation density was defined across the entire genome based on permutations of all the data and which can be modified based on assumptions about any non-biased insertion preferences of MLV, such as the tendency to insertion near transcription start sites of genes. This approach allows one to discover CIS defined by a peak of insertions that exceed an alpha value for significance without suffering from high false positive rates dues to large numbers of insertions. Since the method is scalable to different intervals, significant CIS defining small or larger intervals can be defined, which is important since it is possible that biologically significant intervals vary depending on the genomic context of a given region. For example, CIS could be defined in large tumor suppressor genes using this method that might be missed if CIS over very large intervals are not sought.

These same researchers have gone on the apply this method for detection of statistically significant pairs of common co-occurrence of insertions (CCIs), which could define pairs of cooperating insertion mutations[12]. This method uses a two-dimensional version of the kernel convolution density estimate method described above. These authors defined 86 significant CCIs using published data, among which were previously noted collaborating oncogenes discovered using MLV mutagenesis, such as *Hoxa9* with *Meis1* and *Myc* with *Pim1* [12]. This is among the most exciting attributes of MLV mutagenesis studies, that is, the ability to detect networks of cooperating oncogenes. Such cooperating oncogenes could reveal targets for therapy. For example, the cooperative relationship between *Pim1* and *Myc* could suggest that transformation by *Myc* oncogenes requires Pim1 kinase activity, a prediction that was recently demonstrated experimentally [82]. Therefore, Pim1 kinase inhibition may be of therapeutic value in the many forms of human cancer that have aberrant Myc activity. Attempts to define CCI are just the beginning of more sophisticated network analyses that should be possible using data from MLV mutagenesis studies.

It seems likely that a network of CCI relationships could be assembled by linking all such associations detected in an MLV mutagenesis screen. Data published by de Ridder et al. already establishes that *Myc* exists at a "node" having cooperative relationships with several other oncogenes [12]. By linking all CCIs into a network one might define complexes or signaling pathways. These authors have also suggested that some genes associated with insertion mutations could be clustered into functional classes or pathways prior to analysis to determine if a gene interacts with any of a member of a functional class or pathway. They detected a potential interaction between *Sox4* and any of a number of cyclin dependent kinase genes in this way [12]. In the future, it would also be highly desirable to link these genetic data with specific phenotypes such as the subtype of leukemia (*e.g.* lymphoid versus myeloid leukemia), disease latency, and gene expression profiling.

We have started to develop methods for analysis of CIS-associated genes and leukemia phenotypes. We find that reiterating Fisher's Exact test can be used to associate insertion mutations at specific loci and phenotypes such as surface immunophenotype. Other analytical approaches should be developed to visualize and analyze a systems network for MLV mutagenesis experiments. One complication of this kind of analysis, and CCI analyses, is that we usually assume that any amplified and sequenced insertion mutation is present in all or nearly all cells of the tumor mass. Thus, we assume that any two insertion mutations are present in the same cell and could potentially be cooperating. Similarly, we assume that if a leukemia clone expresses myeloid surface markers, those cells also harbor a particular insertion mutation that was amplified and sequenced from this sample. However, the insertion mutation in question might have been present only in a rare subclone that expresses lymphoid markers and is distantly or unrelated to the majority clone. One method for getting around this problem is to introduce a specific mutation into the germline of mice, as a transgene, and then using MLV mutagenesis to accelerate disease. Then one can potentially identify insertion mutations more prevalent in the transgenic cohort of leukemias compared to non-transgenic control leukemia populations. Many such "sensitized" MLV mutagenesis screens have been performed and are the most common type underway today.

Sensitized MLV mutagenesis screens were pioneered by Dr. Anton Berns at the Netherlands Cancer Institute. He and his collaborators noticed that among T cell leukemias induced by Moloney MLV certain insertion mutations tended to be present within the same clones [4]. Notably, an association was seen between activation of the *Cmyc* gene and the *Pim1* kinase. Indeed, when MLV was used to accelerate T cell leukemia in *Cmyc* overexpressing transgenic mice, the *Pim1* gene was a frequent target of MLV activation [68]. In a brilliant experiment, the MLV infection was used to accelerate disease in *Cmyc* transgenic, *Pim1–/–* mice and the resultant leukemias now had MLV insertion mutations in the related *Pim2* gene [66]. This experiment revealed that it should be possible to use MLV mutagenesis to isolate genes that play a role downstream of another oncogene, or which can act in a parallel pathway, much the way in which *Drosophila* genetics has been used to define multiple members of a pathway important for some developmental process.

More recent sensitized MLV mutagenesis screens have utilized genetic backgrounds of mice carrying alterations similar to known human leukemia-associated mutations, such as tumor suppressor gene loss or expression of specific fusion oncogene products. Several recurrent chromosomal translations create fusion oncoprotein genes in human myeloid leukemia. Some of these have been engineered into mice using homologous recombination in mouse embryonic stem cells. Perhaps not surprisingly, most of these fusion oncoproteins by themselves cannot produce rapid, highly penetrant leukemia in mice. Therefore, investigators have used MLV mutagenesis in attempts to define genes and pathways that can cooperate with expression of these fusion genes. We have investigated genetic events that can cooperate with loss of the *Nf1* tumor suppressor gene in AML development using MLV mutagenesis. The *Myb* and *Bcl11a* genes, among others, were found to cooperate with loss of the *Nf1* gene [79]. These experiments were done by backcrossing an *Nf1* loss of function allele to BXH-2 strain mice, which develop spontaneous AML due to chronic infection with a typical MLV that is passed from mother to offspring [33]. This is a laborious process, and so alternative MLV have been sought for use in strains of mice that are typically used to generate transgenic mice such as the FVB/n and 129/sv strains. For example, the amphotropic MLV, 4070A, has been used to accelerate AML in mice expressing a *CBFbeta-MYH11* transgene, which mimics the effect of the inv(16) event in human AML [6]. Activation of the *Plg1* and *Plgl2* genes were identified as cooperating genetic events, shown to be overexpressed in some human AML, and capable of cooperating with *CFBbeta-MYH11* in a mouse model [32]. The 4070A virus is unfortunately n-tropic and so not useful in the many laboratory strains of mice that carry the $Fv1^b$ allele, such as C57BL/6 [76]. In addition, 4070A can infect human cells and so safety issues are raised. Dr. Linda Wolff has created a chimeric MLV retaining beneficial properties of Moloney and 4070A MLV [77]. This virus, called MOL4070LTR, can induce AML, like 4070A, or T cell ALL and is nb-tropic, like Moloney MLV, and can therefore be used in most laboratory strains of mice. We have used MOL4070LTR to investigate genetic events that cooperate with a prototypical fusion oncogene involving the *MLL* gene, called *MLL-AF9*. The *MLL* gene encodes a Trithorax-related chromatin protein that seems to be required for expression of specific target genes such as some of the *HOX* genes [9]. While some of the fusion oncogenes encode chimeric transcription factors whose activity may be hard to target pharmacologically, it is possible that MLV mutagenesis will uncover critical cooperating genetic events that can be targeted pharmacologically. The same could be said for loss of tumor suppressor genes.

Recently, Berns and colleagues used MLV mutagenesis to investigate genetic events that could cooperate with loss of *p53* or $p19^{Arf}$ function in T cell leukemia genesis [63]. This work is notable for several reasons. The first being that p53 and p19Arf had often been assumed to suppress tumor formation by acting in the same pathway only, and so one might expect identical cooperating events would be identified. However, different cooperating genetic events were found in the two groups. The work also demonstrated the power of using very large numbers of leukemia specimens, several hundred, and using high throughput amplification and sequencing methods to identify MLV insertion mutations. Certainly, greater numbers of genes can be implicated on cancer development by doing so.

Other desirable sensitized, or context-specific, genetic screens can be envisioned using MLV mutagenesis. Of course, the issue of multi-clonality described above complicates all studies in which the tumor genotype (in this case specific insertion mutations) is to be associated with specific phenotypes (e.g., gene expression patterns, surface marker expression) or behaviors (e.g., response to chemotherapy). However, this issue can be ameliorated in part by the study of very large data sets.

Nevertheless, it would be ideal to unequivocally separate all genetically distinct tumor clones prior to any type of analysis. This issue was reviewed recently by Kool and Berns [30]. Methods to do this could involve creating permanent cell lines from MLV-induced leukemias, passaging limited numbers of leukemia cells in mice, or developing methods for analysis of insertions and gene expression patterns from one or very few cells in a section of an MLV-induced leukemia. None of these is a perfect solution to the problem, but some attempt should be made to more reliably allow one to correlate leukemia phenotype with insertion mutation genotype. MLV mutagenesis studies have been most often done in order to discover new leukemia genes and pathways. In the future, it would be desirable to be able to associate specific insertion mutations with specific clinically relevant traits, such as chemotherapy resistance, disease aggressiveness or ability to synergize with specific germline mutations or exposure histories. Several examples of work such as this are described below.

Using MLV-accelerated, *Nf1*-deficient, primary AML samples, Lauchle et al., were able to select in vivo for AML cells resistant to a MEK inhibitor [36]. Interestingly, the authors were able to find MEK inhibitor resistant forms of the primary AMLs that harbored insertions affecting *Rasgrp1*, *Rasgrp4* and *Mapk14*. The authors found that expression of the *Rasgrp* genes at high levels, or downregulation of *Mapk14*, could induce MEK inhibitor resistance. Another group used MLV mutagenesis to select for lymphoma cells with enhanced ability to invade and crawl underneath a monolayer of fibroblasts in culture [23]. This led to the identification of the *Tiam1* gene, encoding a guanine nucleotide exchange factor (GEF) for Rac GTPases, helping to establish a role for Rac proteins in cell migration and tumor dispersal. Many similar studies should be undertaken. In general, models of MLV-induced leukemia and lymphoma have been underutilized for treatment studies using conventional chemotherapy or molecularly targeted therapies. Indeed, mouse modeling of human cancer has been mostly limited to the study of cancer as it might first present, but not as it exists in patients after responding to chemotherapy and/or radiation. It will be very interesting to examine the evolution of a transformed clone in response to selective pressure applied by therapeutic agents. Such studies could allow the identification of resistance mechanisms and/or define the genetic subtypes that will be most sensitive to a given treatment.

2.4 Comparative Oncogenomics

Once the screen for cancer-related genes has been undertaken using MLV mutagenesis and significant CIS-associated genes have been identified, many challenges remain. Among the most challenging is interpreting the results of viral insertional mutagenesis screens to determine what the results have to say about the process of cancer development in humans. Insertional mutagenesis screens have led to the identification of genes that play important biological roles in the normal development of target tissues. However, the prime motivation of these studies is the identification of novel human cancer genes and pathways. A second motivation, only recently studied on a large scale, is the identification of specific patterns of concurrent gene mutation. It can be challenging to achieve these goals because the corresponding tumor type which should be examined in human patients is not always clear, based only on the MLV-induced leukemias. Also, the MLV leukemias may not recapitulate certain types of mutagenic alterations that are common in human leukemia. For example, the *Ras* genes are not common targets of MLV insertional mutagenesis, but are frequently mutated by activating point mutations in specific codons in human leukemias. It is possible that *Ras* genes are not identified at CIS because *Ras* gene overexpression is not sufficient to induce a strong enough transforming signal to be selected for in hematopoietic cells. However, the guanine nucleotide exchange factors, such as *Rasgrp1*, are frequently targeted at CIS [16, 28, 37]. These GEFs activate Ras proteins downstream and can transform cells simply by overexpression [16, 29, 53, 56]. So, in addition to considering the specific genes associated with CIS, one must consider the pathways revealed by MLV insertional mutagenesis screens. Nevertheless, certain examples in which CIS-associated target genes play a direct role in human leukemia as targets of mutation deserve mention.

Several important transcription factors involved in leukemia development have been identified at common sites of MLV integration. The *Myc* gene itself was identified both as an acute transforming viral oncogene in an acute transforming retrovirus from chicken and murine retroviruses and as a recurrent target of MLV integration [70]. The *Myb* gene is a well known target of insertional mutagenesis in myeloid and lymphoid leukemia [28, 44, 76]. This transcription factor is required for normal hematopoiesis [45], and was long thought to not play a direct role in human leukemia. However, recent studies show that these conclusions were premature as *MYB* is often amplified and overexpressed in human leukemia [8, 31, 46]. These studies show that even well known CIS associated genes should be carefully examined for their potential role in human leukemia, in light of new genome wide approaches for characterizing cancer genomes. A large public catalog of CIS-associated genes identified by MLV mutagenesis studies is maintained by Dr. Keiko Agaki, now at the Ohio State University, which lists over 300 CIS genes [1]. Although recent analyses suggest that many of these may not be statistically significant [13, 78] given the fact that MLV do not actually randomly integrate into the genome, this list is still highly likely to contain many loci directly involved in human leukemia via the acquisition of somatically acquired genetic or epigenetic changes. For example, studies on the *Evi2* CIS helped identify the human gene for Neurofibromatosis Type 1 (NF1), a tumor suppressor gene that causes cancer susceptibility when inherited in a mutant form [52]. The fact that human *HOX* genes can be direct leukemia genes was made possible by studies of BXH-2 strain MLV-induced AMLs [49, 50]. The HOX co-factor *Meis1* was discovered as a *Hoxa9* cooperating AML oncogene in studies of the same strain. *Hoxa* and *Meis1* gene activity characterizes the large subset of human leukemias with *MLL* gene rearrangements and the activity of these proteins seems to play a major role in transformation by MLL fusion oncoproteins [2]. This finding was an important outcome of the MLV mutagenesis studies done in BXH-2 strain mice and subsequent validation studies.

Evidence that MLV screens can identify relevant pathways for human leukemia has been obtained by large-scale comparative studies using gene expression profiling of human leukemia genes also. Scientists at Erasmus University in Rotterdam, The Netherlands, have compared CIS-associated genes with gene expression profiling from human AML [18]. Their results show that genes near CIS from the RTCDG database are very significantly enriched for those that are differentially expressed among human AML subtypes, a feature not seen for genes farther away from CIS. Therefore, the expression of these genes may underlie specific programs of leukemic transformation, perhaps downstream of certain well-known fusion oncoproteins. Their level of expression could also be determinants of the varying clinical behavior of different cases of leukemia.

Much attention has recently been paid to the role of microRNA genes in human cancer development. Not surprisingly, MLV mutagenesis has recently been shown to activate miRNA genes in leukemia also [10, 59, 71]. It seems possible that MLV mutagenesis could also pinpoint other types of encoded genomic information that, when altered, could cause cancer. Such encoded information could include other non-protein encoding RNA species or CIS regulatory elements. As more is learned about the architecture of the genome, the results of CIS mutagenesis projects will need to be re-evaluated.

2.5 Validation of Candidates

Perhaps one of the most difficult challenges faced by an investigator once a large number of CIS associated genes has been identified in an MLV mutagenesis study is to decide which of these genes deserve further study. Two central questions come to mind: Which of these genes can play a role in cellular transformation when altered? And which of the genes, or pathways they act in, are directly involved in human cancer development? The advent of human whole cancer genome re-sequencing, methylation arrays, and extensive mRNA and miRNA expression profiling make it possible to address the second question. However, the answer to the first question remains a challenge for the field of cancer genomics in general.

Despite the seemingly daunting task of studying the function of hundreds of genes in cultured cells, living mice, or other model organisms such as the zebrafish, there are now examples that we can point to as evidence of the practicality of such approaches. One method that is finding widespread utility in the study of cancer genes in maintenance of transformation is based on the use of RNAi [11, 17, 73]. The set of CIS-associated genes identified in a screen can be tested for their role in tumor cell maintenance using cultured leukemia cell lines. However, it is likely that some genes recovered in MLV screens are tumor suppressor genes and would thus not score in such an assay. So for this reason, the validation of some genes may require that they be overexpressed from cDNA expression vectors in cell lines that are partially, but not fully, transformed. While this is conceptually straightforward, the choice of such a cell line is not a simple matter. However, large scale projects to produce useful cDNA vectors for all genes in the human and mouse genomes are underway [48]. The projects will of course be very valuable assets in the goal of validating cancer genes and placing them in known signaling pathways. In any case, a role in transformation is not the same as a role in leukemia cell maintenance. Moreover, the effects of knocking down expression of any one gene may have a context dependent effect, and so the phenotype under examination may only be revealed in cell lines that provide the appropriate context. Thus it would perhaps be best to use cell lines derived from the MLV model from which the candidate gene was identified. Finally, some oncogenes probably make a contribution to leukemia development that is not easily revealed in a cell culture assay designed to measure a decrease in cell proliferation or viability. Such phenotypes could have to do with self-renewal, differentiation, interaction with non-tumor cells (i.e., other cells of the immune system), or the response to endogenous growth factors which may not be present in the cell culture system. For this reason, the closer the assay is to true leukemia formation in vivo, the better.

The well known bone marrow transduction and transplantation (BMTT) assay would seem to be ideally suited to validating candidate oncogenes from MLV mutagenesis screens and indeed the assay has been proposed for this use [14], and used by many for such a purpose [79]. However, the assay is technically demanding and the generation of 100 or more high titer MLV based retroviral pools for a project like this is not a simple matter. Even so, Dr. Scott Lowe's lab has succeeded in using pools of large numbers of retroviral shRNA vectors to screen for genes that, when "knocked down" in p53-deficient hepatoblasts, can induce hepatocellular carcinoma in mice [81]. This and other projects make it clear that retroviral transduction studies are a useful method for validating cancer genes in vivo in mice.

Other approaches to validate candidate cancer genes in transgenic mice may be more flexible and powerful. One creative method, developed by Dr. Pentao Liu, involves cloning candidate oncogenes into transposon vectors in which the cDNA can only be expressed after mobilization into an actively transcribed gene [61]. By creating mice carrying many such vectors with a variety of oncogenes and mobilizing them tissue-specifically to induce tumors, one can determine which have been activated by insertion into an active gene. This method has the advantage of allowing sets of candidate oncogenes to be simultaneously tested in a variety of tissues using a Cre/LoxP-regulated transposase transgene and allows for the possibility that only a certain combination of oncogenes might collaborate to induce cancer. While clever, this approach still requires the production of gene-trap cDNA vectors and a transgenic line of mice. It is also still unclear how many different oncogene candidates would be practical to include in a single transgenic mouse strain. In any case, this technology is emblematic of the kind of new thinking that should be undertaken in projects aimed at in vivo functional validation of oncogene candidates.

One approach is to use transposon-based methods of gene delivery, which use transfected plasmids, rather than retroviral vectors. This approach has the advantage that one need only produce plasmid expression vectors for each cDNA or shRNA to be tested. We, and others, have shown that the *Sleeping Beauty* transposon system can be used to deliver of cDNAs and shRNAs to cause tumors in mice when delivered along with a source of the *Sleeping Beauty* transposase [5, 27, 72]. While these were solid tumors, recent data shows that it is possible to use *Sleeping Beauty* to efficiently deliver expressed cDNAs to early hematopoietic precursors via electroporation [41]. This approach has strong potential because cDNA delivery via transposition also allows efficient co-delivery of multiple vectors to the same cell in vivo, as has been demonstrated for glioma induced in mice [75].

2.6 Do Murine Retroviruses Cause Human Cancer?

Recent studies have indicated that a class of xenotropic murine retroviruses (XMRV) are present in some human tissue samples [15, 65]. This virus is closely related to the MLV. XMRV has been linked to both prostate cancer and chronic fatigue development [38]. Interestingly, some reports suggest that XMRV infection is present in prostate tumors, but only common in prostate tumors from patients with a polymorphism in the *Rnasel* gene, which seems to render them susceptible to viral infections [65]. Controversy surrounds the question of how XMRV may contribute to prostate cancer development. Some reports show that XMRV sequences are not common in human prostate cancer samples [24] or lymphocytes from chronic fatigue syndrome patients [19]. Other reports suggest that the epithelial tumor cells themselves are infected with XMRV and that among the insertion sites are genes with roles in growth regulation [58]. It is therefore unclear whether XMRV infection might contribute to prostate cancer development by an insertional mutagenesis mechanism. If XMRV can contribute to human disease, the implications are profound and the lessons learned from the study of MLV-induced leukemia and lymphoma in rodents will be useful in interpreting the situation with XMRV.

2.7 Summary and Future Perspectives

The future of MLV mutagenesis studies promises to be very exciting. It should involve selection for novel and more specific phenotypes. The history of *Drosophila* genetics shows that screens for more specific phenotypes was the key to unraveling specific and critical pathways in developmental processes. Although difficult to design and implement, screens for such specific subtypes of leukemia, or leukemic phenotypes, are the best way to generate the most useful lists of CIS-associated genes. Insertion mutations that allow survival and expansion in a specific setting would seem to offer the best chance to understand how the leukemia phenotype evolves and is maintained. However, several problems must be addressed if MLV mutagenesis is to achieve this goal. This includes the best definition of a significant CIS and co-occurring CIS, methods to ensure saturation recovery of MLV insertion mutations, and identification of the targeted gene(s) at each CIS. Moreover, methods to deal with the clonal heterogeneity of MLV-induced leukemias, which complicates associations of genotype with phenotype, must be addressed. The best methods for comparisons to human leukemia genome alterations must be sought. Finally, new higher throughput methods for functional validation of CIS-associated genes should be developed so that more candidates can be tested in a reasonable period of time.

References

1. Akagi, K., Suzuki, T., Stephens, R. M., Jenkins, N. A., & Copeland, N. G. (2004). RTCGD: Retroviral tagged cancer gene database. *Nucleic Acids Research*, *32*, D523–D527.
2. Ayton, P. M., & Cleary, M. L. (2003). Transformation of myeloid progenitors by MLL oncoproteins is dependent on Hoxa7 and Hoxa9. *Genes Dev, 17*, 2298–2307.
3. Bedigian, H. G., Taylor, B. A., & Meier, H. (1981). Expression of murine leukemia viruses in the highly lymphomatous BXH-2 recombinant inbred mouse strain. *Journal of Virology*, *39*, 632–640.
4. Berns, A. (1988). Provirus tagging as an instrument to identify oncogenes and to establish synergism between oncogenes. *Archives of Virology*, *102*, 1–18.
5. Carlson, C. M., Frandsen, J. L., Kirchhof, N., McIvor, R. S., & Largaespada, D. A. (2005). Somatic integration of an oncogene-harboring Sleeping Beauty transposon models liver tumor development in the mouse. *Proceedings of the National Academy of Sciences of the USA*, *102*, 17059–17064.
6. Castilla, L. H., Perrat, P., Martinez, N. J., Landrette, S. F., Keys, R., Oikemus, S., et al. (2004). Identification of genes that synergize with Cbfb-MYH11 in the pathogenesis of acute myeloid leukemia. *Proceedings of the National Academy of Sciences of the USA*, *101*, 4924–4929.
7. Caudell, D., Harper, D. P., Novak, R. L., Pierce, R. M., Slape, C., Wolff, L., et al. Retroviral insertional mutagenesis identifies Zeb2 activation as a novel leukemogenic collaborating event in CALM-AF10 transgenic mice. *Blood*, *115*, 1194–1203.
8. Clappier, E., Cuccuini, W., Kalota, A., Crinquette, A., Cayuela, J. M., Dik, W. A., et al. (2007). The C-MYB locus is involved in chromosomal translocation and genomic duplications in human T-cell acute leukemia (T-ALL), the translocation defining a new T-ALL subtype in very young children. *Blood*, *110*, 1251–1261.
9. Collins, E. C., & Rabbitts, T. H. (2002). The promiscuous MLL gene links chromosomal translocations to cellular differentiation and tumour tropism. *Trends in Molecular Medicine*, *8*, 436–442.
10. Cui, J. W., Li, Y. J., Sarkar, A., Brown, J., Tan, Y. H., Premyslova, M., et al. (2007). Retroviral insertional activation of the Fli-3 locus in erythroleukemias encoding a cluster of microRNAs that convert Epo-induced differentiation to proliferation. *Blood*, *110*, 2631–2640.
11. DasGupta, R., Kaykas, A., Moon, R. T., & Perrimon, N. (2005). Functional genomic analysis of the Wnt-wingless signaling pathway, *308*, 826–833 (Epub 2005 Apr 2007).
12. de Ridder, J., Kool, J., Uren, A., Bot, J., Wessels, L., & Reinders, M. (2007). Co-occurrence analysis of insertional mutagenesis data reveals cooperating oncogenes. *Bioinformatics*, *23*, i133–i141.
13. de Ridder, J., Uren, A., Kool, J., Reinders, M., & Wessels, L. (2006). Detecting statistically significant common insertion sites in retroviral insertional mutagenesis screens. *PLoS Computationl Biology*, *2*, e166.
14. DePinho, R. A., & Jacks, T. (1999). A bumper crop of cancer genes. *Nature Genetics*, *23*, 253–254.
15. Dong, B., Kim, S., Hong, S., Das Gupta, J., Malathi, K., Klein, E. A., et al. (2007). An infectious retrovirus susceptible to an IFN antiviral pathway from human prostate tumors. *Proceedings of the National Academy of Sciences of the USA*, *104*, 1655–1660.

16. Dupuy, A. J., Morgan, K., von Lintig, F. C., Shen, H., Acar, H., Hasz, D. E., et al. (2001). Activation of the Rap1 guanine nucleotide exchange gene, CalDAG-GEF I, in BXH2 murine myeloid leukemia. *Journal of Biological Chemistry*, *22*, 22.
17. Ebert, B. L., Pretz, J., Bosco, J., Chang, C. Y., Tamayo, P., Galili, N., et al. (2008). Identification of RPS14 as a 5q- syndrome gene by RNA interference screen. *Nature*, *451*, 335–339.
18. Erkeland, S. J., Verhaak, R. G., Valk, P. J., Delwel, R., Lowenberg, B., & Touw, I. P. (2006). Significance of murine retroviral mutagenesis for identification of disease genes in human acute myeloid leukemia. *Cancer Research*, *66*, 622–626.
19. Erlwein, O., Kaye, S., McClure, M. O., Weber, J., Wills, G., Collier, D., et al. Failure to detect the novel retrovirus XMRV in chronic fatigue syndrome. *PLoS One*, *5*, e8519.
20. Fan, H. (1997). Leukemogenesis by Moloney murine leukemia virus: A multistep process. *Trends in Microbiology*, *5*, 74–82.
21. Fan, H., Mittal, S., Chute, H., Chao, E., & Pattengale, P. K. (1986). Rearrangements and insertions in the Moloney murine leukemia virus long terminal repeat alter biological properties in vivo and in vitro. *Journal of Virology*, *60*, 204–214.
22. Gardner, M. B. (1978). Type C viruses of wild mice: Characterization and natural history of amphotropic, ecotropic, and xenotropic MuLv. *Current Topics in Microbiology and Immunology*, *79*, 215–259.
23. Habets, G. G., Scholtes, E. H., Zuydgeest, D., van der Kammen, R. A., Stam, J. C., Berns, A., et al. (1994). Identification of an invasion-inducing gene, Tiam-1, that encodes a protein with homology to GDP-GTP exchangers for Rho-like proteins. *Cell*, *77*, 537–549.
24. Hohn, O., Krause, H., Barbarotto, P., Niederstadt, L., Beimforde, N., Denner, J., et al. (2009). Lack of evidence for xenotropic murine leukemia virus-related virus(XMRV) in German prostate cancer patients. *Retrovirology*, *6*, 92.
25. Jenkins, N. A., Copeland, N. G., Taylor, B. A., Bedigian, H. G., & Lee, B. K. (1982). Ecotropic murine leukemia virus DNA content of normal and lymphomatous tissues of BXH-2 recombinant inbred mice. *Journal of Virology*, *42*, 379–388.
26. Jin, S., Zhao, H., Yi, Y., Nakata, Y., Kalota, A., & Gewirtz, A. M. (2010). c-Myb binds MLL through menin in human leukemia cells and is an important driver of MLL-associated leukemogenesis. *Journal of Clinical Investigation*, *120*, 593–606.
27. Keng, V. W., Villanueva, A., Chiang, D. Y., Dupuy, A. J., Ryan, B. J., Matise, I., et al. (2009). A conditional transposon-based insertional mutagenesis screen for genes associated with mouse hepatocellular carcinoma. *Nature Biotechnology*, *27*, 264–274.
28. Kim, R., Trubetskoy, A., Suzuki, T., Jenkins, N. A., Copeland, N. G., & Lenz, J. (2003). Genome-based identification of cancer genes by proviral tagging in mouse retrovirus-induced T-cell lymphomas. *Journal of Virology*, *77*, 2056–2062.
29. Klinger, M. B., Guilbault, B., Goulding, R. E., & Kay, R. J. (2005). Deregulated expression of RasGRP1 initiates thymic lymphomagenesis independently of T-cell receptors. *Oncogene*, *24*, 2695–2704.
30. Kool, J., & Berns, A. (2009). High-throughput insertional mutagenesis screens in mice to identify oncogenic networks. *Nature Review Cancer*, *9*, 389–399.
31. Lahortiga, I., De Keersmaecker, K., Van Vlierberghe, P., Graux, C., Cauwelier, B., Lambert, F., et al. (2007). Duplication of the MYB oncogene in T cell acute lymphoblastic leukemia. *Nature Genetics*, *39*, 593–595.
32. Landrette, S. F., Kuo, Y. H., Hensen, K., S., Perrat, P. N., Van de Ven, W. J., et al. (2005). Plag1 and Plagl2 are oncogenes that induce acute myeloid leukemia in cooperation with Cbfb-MYH11. *Blood*, *105*, 2900–2907.
33. Largaespada, D. A., Brannan, C. I., Jenkins, N. A., & Copeland, N. G. (1996). Nf1 deficiency causes Ras-mediated granulocyte/macrophage colony stimulating factor hypersensitivity and chronic myeloid leukaemia. *Nature Genetics*, *12*, 137–143.
34. Largaespada, D. A., & Collier, L. S. (2008). Transposon-mediated mutagenesis in somatic cells: Identification of transposon-genomic DNA junctions. *Methods in Molecular Biology*, *435*, 95–108.

35. Largaespada, D. A., Shaughnessy, J. D., Jenkins, N. A., & Copeland, N. G. (1995). Retroviral insertion at the Evi-2 locus in BXH-2 myeloid leukemia cell lines disrupts Nf1 expression without changes in steady state ras-GTP levels. *Journal of Virology*, *69*, 5095–5102.
36. Lauchle, J. O., Kim, D., Le, D. T., Akagi, K., Crone, M., Krisman, K., et al. (2009). Response and resistance to MEK inhibition in leukaemias initiated by hyperactive Ras. *Nature*, *461*, 411–414.
37. Li, J., Shen, H., Himmel, K. L., Dupuy, A. J., Largaespada, D. A., Nakamura, T., et al. (1999). Leukaemia disease genes: Large-scale cloning and pathway predictions [see comments]. *Nature Genetics*, *23*, 348–353.
38. Lombardi, V. C., Ruscetti, F. W., Das Gupta, J., Pfost, M. A., Hagen, K. S., Peterson, D. L., et al. (2009). Detection of an infectious retrovirus, XMRV, in blood cells of patients with chronic fatigue syndrome. *Science*, *326*, 585–589.
39. Lund, A. H., Turner, G., Trubetskoy, A., Verhoeven, E., Wientjens, E., Hulsman, D., et al. (2002). Genome-wide retroviral insertional tagging of genes involved in cancer in Cdkn2a-deficient mice. *Nature Genetics*, *32*, 160–165.
40. Maeda, N., Fan, H., & Yoshikai, Y. (2008). Oncogenesis by retroviruses: Old and new paradigms. *Reviews in Medical Virology*, *18*, 387–405.
41. Mates, L., Chuah, M. K., Belay, E., Jerchow, B., Manoj, N., Acosta-Sanchez, A., et al. (2009). Molecular evolution of a novel hyperactive Sleeping Beauty transposase enables robust stable gene transfer in vertebrates. *Nature Genetics*, *41*, 753–761.
42. Mikkers, H., Allen, J., Knipscheer, P., Romeijn, L., Hart, A., Vink, E., et al. (2002). High-throughput retroviral tagging to identify components of specific signaling pathways in cancer. *Nature Genetics*, *32*, 153–159.
43. Miller, D. G., Adam, M. A., & Miller, A. D. (1990). Gene transfer by retrovirus vectors occurs only in cells that are actively replicating at the time of infection. *Molecular and Cellular Biology*, *10*, 4239–4242.
44. Mukhopadhyaya, R., & Wolff, L. (1992). New sites of proviral integration associated with murine promonocytic leukemias and evidence for alternate modes of c-myb activation [published erratum appears in *J Virol* 1993 May;67(5):2960]. *Journal of Virology*, *66*, 6035–6044.
45. Mukouyama, Y., Chiba, N., Mucenski, M. L., Satake, M., Miyajima, A., Hara, T., et al. (1999). Hematopoietic cells in cultures of the murine embryonic aorta-gonad-mesonephros region are induced by c-Myb. *Current Biology*, *9*, 833–836.
46. Murati, A., Gervais, C., Carbuccia, N., Finetti, P., Cervera, N., Adelaide, J., et al. (2009). Genome profiling of acute myelomonocytic leukemia: Alteration of the MYB locus in MYST3-linked cases. *Leukemia*, *23*, 85–94.
47. Muriaux, D., & Rein, A. (2003). Encapsidation and transduction of cellular genes by retroviruses. *Frontiers in Bioscience*, *8*, d135–d142.
48. Nagase, T., Yamakawa, H., Tadokoro, S., Nakajima, D., Inoue, S., Yamaguchi, K., et al. (2008). Exploration of human ORFeome: High-throughput preparation of ORF clones and efficient characterization of their protein products. *DNA Research*, *15*, 137–149.
49. Nakamura, T., Largaespada, D. A., Lee, M. P., Johnson, L. A., Ohyashiki, K., Toyama, K., et al. (1996a). Fusion of the nucleoporin gene NUP98 to HOXA9 by the chromosome translocation t(7;11)(p15;p15) in human myeloid leukaemia. *Nature Genetics*, *12*, 154–158.
50. Nakamura, T., Largaespada, D. A., Shaughnessy, J. D., Jr., Jenkins, N. A., & Copeland, N. G. (1996b). Cooperative activation of Hoxa and Pbx1-related genes in murine myeloid leukaemias. *Nature Genetics*, *12*, 149–153.
51. Nishigaki, K., Hanson, C., Thompson, D., Yugawa, T., Hisasue, M., Tsujimoto, H., et al. (2002). Analysis of the disease potential of a recombinant retrovirus containing Friend murine leukemia virus sequences and a unique long terminal repeat from feline leukemia virus. *Journal of Virology*, *76*, 1527–1532.

52. O'Connell, P., Viskochil, D., Buchberg, A. M., Fountain, J., Cawthon, R. M., Culver, M., et al. (1990). The human homolog of murine Evi-2 lies between two von Recklinghausen neurofibromatosis translocations. *Genomics*, *7*, 547–554.
53. Oki-Idouchi, C. E., & Lorenzo, P. S. (2007). Transgenic overexpression of RasGRP1 in mouse epidermis results in spontaneous tumors of the skin. *Cancer Research*, *67*, 276–280.
54. Ou, C. Y., Boone, L. R., Koh, C. K., Tennant, R. W., & Yang, W. K. (1983). Nucleotide sequences of gag-pol regions that determine the Fv-1 host range property of BALB/c N-tropic and B-tropic murine leukemia viruses. *Journal of Virology*, *48*, 779–784.
55. Patriotis, C., Makris, A., Bear, S. E., & Tsichlis, P. N. (1993). Tumor progression locus 2 (Tpl-2) encodes a protein kinase involved in the progression of rodent T-cell lymphomas and in T-cell activation. *Proceedings of the National Academy of Sciences of the USA*, *90*, 2251–2255.
56. Reuther, G. W., Lambert, Q. T., Rebhun, J. F., Caligiuri, M. A., Quilliam, L. A., & Der, C. J. (2002). RasGRP4 is a novel Ras activator isolated from acute myeloid leukemia. *Journal of Biological Chemistry*, *277*, 30508–30514.
57. Risser, R., Horowitz, J. M., & McCubrey, J. (1983). Endogenous mouse leukemia viruses. *Annual Review of Genetics*, *17*, 85–121.
58. Schlaberg, R., Choe, D. J., Brown, K. R., Thaker, H. M., & Singh, I. R. (2009). XMRV is present in malignant prostatic epithelium and is associated with prostate cancer, especially high-grade tumors. *Proceedings of the National Academy of Sciences of the USA*, *106*, 16351–16356.
59. Slape, C., Hartung, H., Lin, Y. W., Bies, J., Wolff, L., & Aplan, P. D. (2007). Retroviral insertional mutagenesis identifies genes that collaborate with NUP98-HOXD13 during leukemic transformation. *Cancer Research*, *67*, 5148–5155.
60. Starr, T. K., & Largaespada, D. A. (2005). Cancer gene discovery using the sleeping beauty transposon. *Cell Cycle*, *4*, 1744–1748.
61. Su, Q., Prosser, H. M., Campos, L. S., Ortiz, M., Nakamura, T., Warren, M., et al. (2008). A DNA transposon-based approach to validate oncogenic mutations in the mouse. *Proceedings of the National Academy of Sciences of the USA*, *105*, 19904–19909.
62. Turcotte, K., Gauthier, S., Tuite, A., Mullick, A., Malo, D., & Gros, P. (2005). A mutation in the Icsbp1 gene causes susceptibility to infection and a chronic myeloid leukemia-like syndrome in BXH-2 mice. *Journal of Experimental Medicine*, *201*, 881–890.
63. Uren, A. G., Kool, J., Matentzoglu, K., de Ridder, J., Mattison, J., van Uitert, M., et al. (2008). Large-scale mutagenesis in p19(ARF)- and p53-deficient mice identifies cancer genes and their collaborative networks. *Cell*, *133*, 727–741.
64. Uren, A. G., Mikkers, H., Kool, J., van der Weyden, L., Lund, A. H., Wilson, C. H., et al. (2009). A high-throughput splinkerette-PCR method for the isolation and sequencing of retroviral insertion sites. *Nature Protocols*, *4*, 789–798.
65. Urisman, A., Molinaro, R. J., Fischer, N., Plummer, S. J., Casey, G., Klein, E. A., et al. (2006). Identification of a novel Gammaretrovirus in prostate tumors of patients homozygous for R462Q RNASEL variant. *PLoS Pathogens*, *2*, e25.
66. van der Lugt, N. M., Domen, J., Verhoeven, E., Linders, K., van der Gulden, H., Allen, J., et al. (1995). Proviral tagging in E mu-myc transgenic mice lacking the Pim-1 proto-oncogene leads to compensatory activation of Pim-2. *EMBO Journal*, *14*, 2536–2544.
67. van Lohuizen, M., Breuer, M., & Berns, A. (1989). N-myc is frequently activated by proviral insertion in MuLV-induced T cell lymphomas. *EMBO Journal*, *8*, 133–136.
68. van Lohuizen, M., Verbeek, S., Scheijen, B., Wientjens, E., van der Gulden, H., & Berns, A. (1991). Identification of cooperating oncogenes in E mu-myc transgenic mice by provirus tagging [see comments]. *Cell*, *65*, 737–752.
69. van Ooyen, A., & Nusse, R. (1984). Structure and nucleotide sequence of the putative mammary oncogene int-1; proviral insertions leave the protein-encoding domain intact. *Cell*, *39*, 233–240.
70. Varmus, H. E. (1983). Using retroviruses as insertional mutagens to identify cellular oncogenes. *Progress in Clinical and Biological Research*, *119*, 23–35.

71. Wang, C. L., Wang, B. B., Bartha, G., Li, L., Channa, N., Klinger, M., et al. (2006). Activation of an oncogenic microRNA cistron by provirus integration. *Proceedings of the National Academy of Sciences of the USA*, *103*, 18680–18684.
72. Wangensteen, K. J., Wilber, A., Keng, V. W., He, Z., Matise, I., Wangensteen, L., et al. (2008). A facile method for somatic, lifelong manipulation of multiple genes in the mouse liver. *Hepatology*, *47*, 1714–1724.
73. Westbrook, T. F., Martin, E. S., Schlabach, M. R., Leng, Y., Liang, A. C., Feng, B., et al. (2005). A genetic screen for candidate tumor suppressors identifies REST. *Cell*, *121*, 837–848.
74. Whitcomb, J. M., & Hughes, S. H. (1992). Retroviral reverse transcription and integration: Progress and problems. *Annual Review of Cell Biology*, *8*, 275–306.
75. Wiesner, S. M., Decker, S. A., Larson, J. D., Ericson, K., Forster, C., Gallardo, J. L., et al. (2009). De novo induction of genetically engineered brain tumors in mice using plasmid DNA. *Cancer Research*, *69*, 431–439.
76. Wolff, L., Koller, R., & Davidson, W. (1991). Acute myeloid leukemia induction by amphotropic murine retrovirus (4070A): Clonal integrations involve c-myb in some but not all leukemias. *Journal of Virology*, *65*, 3607–3616.
77. Wolff, L., Koller, R., Hu, X., & Anver, M. R. (2003). A Moloney murine leukemia virus-based retrovirus with 4070A long terminal repeat sequences induces a high incidence of myeloid as well as lymphoid neoplasms. *Journal of Virology*, *77*, 4965–4971.
78. Wu, X., Luke, B. T., & Burgess, S. M. (2006). Redefining the common insertion site. *Virology*, *344*, 292–295.
79. Yin, B., Delwel, R., Valk, P. J., Wallace, M. R., Loh, M. L., Shannon, K. M., et al. (2009). A retroviral mutagenesis screen reveals strong cooperation between Bcl11a overexpression and loss of the Nf1 tumor suppressor gene. *Blood*, *113*, 1075–1085.
80. Yin, B., & Largaespada, D. A. (2007). PCR-based procedures to isolate insertion sites of DNA elements. *Biotechniques*, *43*, 79–84.
81. Zender, L., Xue, W., Zuber, J., Semighini, C. P., Krasnitz, A., Ma, B., et al. (2008). An oncogenomics-based in vivo RNAi screen identifies tumor suppressors in liver cancer. *Cell*, *135*, 852–864.
82. Zippo, A., De Robertis, A., Serafini, R., & Oliviero, S. (2007). PIM1-dependent phosphorylation of histone H3 at serine 10 is required for MYC-dependent transcriptional activation and oncogenic transformation. *Nature Cell Biology*, *9*, 932–944.

Chapter 3
Gene Discovery by MMTV Mediated Insertional Mutagenesis

Annabel Vendel-Zwaagstra and John Hilkens

3.1 Introduction

Transformation of a normal cell into a cancer cell requires sequential accumulation of several genetic changes that affect various collaborating signaling cascades mainly involved in cell proliferation, survival and development [46]. To develop novel specific therapeutic compounds for cancer treatment, identification of these cancer causing genes, and subsequently the oncogenic pathways in which these genes act, is of utmost importance. Although a great deal of insight in the cellular mechanisms that lead to cancer has been obtained, many key players are still unidentified. Retroviral insertional mutagenesis (IM) screens provide one of the most efficient tools to identify genes involved in tumorigenesis and from there the specific oncogenic pathways involved. Retroviruses integrate their proviral DNA into the host genome as part of their replication cycle. Integration of the provirus in the genomic DNA of the host cell can lead to inappropriate activation of flanking genes or other mutational events that can cause oncogenic transformation of the infected cells. The genes affected by the inserted provirus can be predicted by determining the chromosomal localization of the proviral DNA (reviewed by [51, 62, 145, 147]. Although insertional mutagenesis is less suitable to trace tumor suppressor genes, it is particularly suited for the discovery of oncogenes.

In the western world, breast cancer is the most prevailing cancer among women. In some countries, one in eight women may contract this disease and approximately 25% will die from it [59]. Several genes have been shown to be involved in human breast cancer, including *ERBB2/Neu/Her2*, *ERBB1/EGFR*, *PIK3CA*, *MYC*, *CCND1*, *P53*, *PTEN*, *CDH1*, *RB1*, *CDKN2A(p14)*, *BRCA1* and *BRCA2* (selected from the Cancer Gene Census and Catalogue of Somatic Mutations in Cancer (COSMIC) databases). To identify other important genes involved in human breast cancer development in a systematic way, a detour to mouse models for breast cancer is

J. Hilkens (✉)
Divisions of Molecular Genetics, The Netherlands Cancer Institute, 1066 CX Amsterdam, The Netherlands
e-mail: j.hilkens@nki.nl

A.J. Dupuy, D.A. Largaespada (eds.), *Insertional Mutagenesis Strategies in Cancer Genetics*, DOI 10.1007/978-1-4419-7656-7_3,

often being made. Although the frequency of specific tumor subtypes and specific oncogenes involved may vary considerably in mouse and man, most of the known oncogenic pathways leading to tumors in mice also contribute to oncogenesis in man. Despite the potential difficulties to determine the relevance of the results obtained in mouse models for human cancer, mouse models provide the best available experimental tools to discover genetic pathways altered in breast cancer. The relevance IM for human tumorigenesis can be further enhanced by performing IM in mouse models in which activated known (human) oncogenes are introduced or certain tumor suppressor genes are inactivated. Depending on the oncogene or tumor suppressor gene, tumor type and subtypes that develop in these mice mimic to a very high degree those occurring in humans [31, 63].

To discover novel oncogenes in mouse mammary tumors by insertional mutagenesis Mouse Mammary Tumor Virus (MMTV) is still frequently used [12, 51, 81, 144], although the use of the recently developed gene tagging technology by transposons could be used as well [20] and Chapter 5. Already more than 25 years ago, the first genes affected by MMTV proviral insertions in mouse mammary tumors were discovered by using very laborious technologies, such as genomic library screening, and Southern blotting. Mainly the more frequently tagged genes were found, almost all belonging to the Wnt and Fgf family. In some strains members of the Notch gene family were also reported as frequent MMTV targets (Table 3.1). Today, the availability of the draft sequence of the mouse genome and high-throughput PCR and parallel sequencing technologies have greatly facilitated the search for proviral integration sites and subsequent identification of retrovirally-activated oncogenes [144]. New sequencing platforms and improved bioinformatics tools will further facilitate high-throughput IM screenings and prediction of MMTV target genes [22]. Moreover, MMTV insertional mutagenesis screens in genetically modified mouse models will identify genes and pathways that collaborate with known cancer genes relevant for human breast cancer [25, 77, 82].

In this chapter, we will first briefly describe some essential features of the biology of the mammary gland, next describe some characteristics of MMTV, in particular properties that are relevant for insertional mutagenesis, followed by a short discussion of MMTV-induced cancers. Subsequently, we will summarize the

Table 3.1 MMTV target genes identified before the genome era

MMTV target gene	References
Wnt1 (int-1)	[110]
Wnt3 (int-4)	[126]
Wnt10b	[77]
Fgf3 (int-2)	[113]
Fgf4	[114]
Fgf8	[82]
Notch1	[25]
Notch4 (int-3)	[35]
eIF3 p48 subunit (int-6)	[84]

MMTV-IM screens performed so far, highlight the major target genes and the pathways activated. Finally, we will focus on the relevance of these target genes and the corresponding pathways for human breast cancer.

3.1.1 Mouse Mammary Gland Biology

The mouse mammary gland is a dynamic organ that after embryonic development undergoes several developmental changes during, puberty, pregnancy, lactation and involution. At embryonic day 11–11.5 five pairs of lens-shaped mammary placodes are formed in the ventral ectoderm in response to signals from the underlying mesenchyme. At day 12.5 the placodal epithelial cells migrate into the dense mammary mesenchyme and form small epithelial buds that slowly increase in size. Wnt and Fgf signalling plays an important role at these stages [18, 156]. At day 15.5 a mammary sprout grows from each placode into the mammary fat pad. By day 16 the sprouts start to branch and a lumen develops. At birth the female mammary gland is composed of a rudimentary ductal tree consisting of the primary duct and 15–20 secondary branches. In response to ovarian steroid hormones during puberty (mainly estrogen) this rudimentary ductal system elongates through proliferation of cap cells in the terminal end buds located at the tips. The elongating ducts bifurcate a few times, which leads to a ductal tree of primary and secondary ducts that fills the entire mammary fat pad. During the recurrent estrous cycles and during pregnancy, lateral buds, also called alveolar buds, form from quiescent cells in the secondary ducts. The lateral buds contain stem cells with the capacity to form either ductal or luminal alveolar cell types. Pregnancy hormones activate rapid proliferation of stem cell populations in the lateral buds and induce an extensive network of tertiary branches that differentiate into secretory alveoli that produce milk that is secreted into the ductal lumens during lactation. Following weaning, massive apoptosis of the lobular alveolar cells takes place resulting in regression of the alveolar structures while stem cells in the alveolar buds await another round of growth and differentiation during subsequent pregnancies [122, 160]. The adult mammary gland is comprised of three cell lineages: myoepithelial cells forming the basal layer of ducts and alveoli, ductal epithelial cells lining the lumen of the ducts, and alveolar epithelial cells that produce the milk proteins and fat during lactation. To generate the entire mammary gland during the next pregnancy, stem cells in the alveolar buds have the capacity of self-renewal and the ability to generate again the three cell lineages that comprise the lactating mammary gland.

3.1.2 Mouse Mammary Tumor Virus

MMTV belongs to the family of Betaretroviridae (B-type retroviruses), which are retroviruses with a distinct morphology such as an eccentric positioned nucleus. Intracytoplasmic A-particles found in MMTV producing cells are the immature,

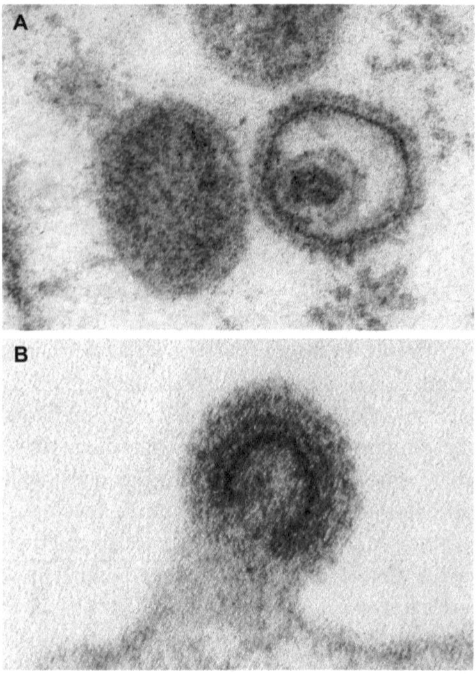

Fig. 3.1 Electron micrographs of different stages of MMTV maturation. (**a**) shows a A-particle (*left*), which is an intracellular immature form of MMTV, and a B-particle (*right*) which is the mature virion. Note the eccentric core of the B-particle, whereas the A-particle does not have a defined core and does not contain a lipid envelope. (**b**) shows a budding virion. The virion is enclosed by the plasma membrane and the core proteins have not matured (H. Jansen and Dr J. Calafat, NKI Amsterdam)

non-enveloped, intracellular precursors of the mature virion (B-particle) (Fig. 3.1). MMTV is a slow transforming retrovirus mainly causing mammary tumors with a latency of 6–12 months. The virus was initially described as an active milk factor that plays an important role in the development of mouse mammary tumors [9]. Besides the exogenous form, which is transmitted through the milk of infected mice from mother to offspring, MMTV can also be transmitted endogenously through stably integrated proviral copies in the germ line. In addition to the mode of transmission and the tropism, regulation of expression of the virus by pregnancy hormones is another important difference between MMTV and other retroviruses [45].

Active infectious mammary tumor virus has only been unequivocally identified in mice. Although sequences related to MMTV (e.g. HERV-K sequences) are present in the human genome and expression of MMTV *env* gene-like sequences have been reported in human breast carcinomas [158, 159], neither an etiological relationship between expression of these sequences and breast cancer nor infectious human MTV particles have been found [6,83]. Also the reported serologic relationships between MMTV and proteins present in human breast cancer cells are highly controversial [38].

In the next paragraphs, we will describe the most important features of MMTV, in particular those that discriminate MMTV from other retroviruses and properties that are relevant for insertional mutagenesis.

3.1.3 Viral Genes and Genomic Organization

The MMTV virion contains two identical single-stranded, non-covalently linked RNA molecules of 8620 nucleotides (nt) [102]. The genomic RNA encodes the *Gag, Dut/Pro, Pol/In,* and *Env* genes and at least two accessory genes, *Sag* and *Rem* (Fig. 3.2). The 5′ end of the genomic RNA contains a 110 nt unique region (U5) and the 3′ end a 1197 nt unique region (U3). Both of these unique regions are respectively preceded and followed by short a direct repeat, R (Fig. 3.2). The R region of MMTV 15 nt and is the smallest reported for known retroviruses, whereas

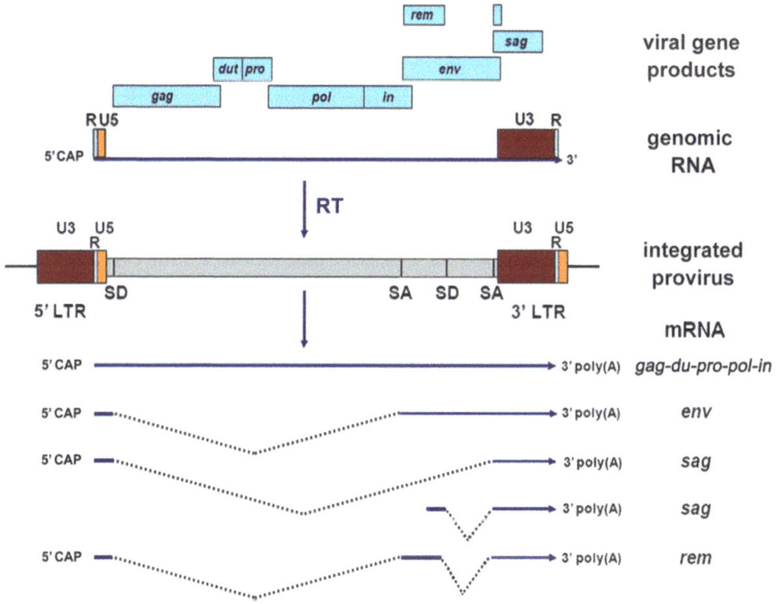

Fig. 3.2 Genomic structure and transcription of MMTV. MMTV proteins are encoded on a polycystronic RNA. The *upper part* of the figure shows the structure of the genomic RNA and the genomic location of the viral gene products. The localization of the R, U3 and U5 regions are indicated on the genomic RNA and the provirus. The entire genomic RNA is 8620 nt. The R region is a 15 nt direct repeat present at the 5′ and 3′ ends of the genomic RNA. The U3 region is ~1200 nt and contains enhancer and promoter sequences but also encodes most of the Sag gene. The U5 region is 110 nt and contains no known protein information. Reversed transcription (RT) activity produces a double stranded DNA molecule, the provirus, in which the U3 and U5 regions are joined and located at both ends of the proviral DNA. Subsequently, the integrase activity catalyses integration of the provirus in the genomic host DNA, as shown in the middle part. The *lower part* of the figure depicts various mRNA products transcribed from the provirus. Except the predominant Sag mRNA, that is produced from an intragenic promoter within the env gene all mRNAs are transcribed from the same start site in the 5′ LTR. As indicated, several viral proteins share part of their coding regions; this is further outlined in the text. Introns are depicted by *v-shaped dotted lines* and exons by *solid lines*. SD: splice donor site; SA: splice acceptor site; RT: reversed transcription. The figure is based on Mertz and colleagues [91] and references cited in the text

the MMTV U3 region is relatively long as compared to other retroviruses. As in other retroviruses, the U3 region contains most elements that regulate transcription of the viral genes. Upon reversed transcription, the long terminal repeats (LTR) are formed at the 5′ and 3′ end of the provirus. The LTR is composed of the R, U3 and U5 regions that are joined as a consequence of the position of the reverse transcriptase primer binding site in the viral RNA and the template switches during proviral DNA synthesis (Fig. 3.2) (for details see [142]).

Figure 3.1 depicts the map of the MMTV proviral DNA and the genes it encodes. The *Gag* gene encodes a polypeptide that is proteolytically processed into the viral core proteins. *Dut* encodes a deoxyuridine triphosphatase (dUTPase). The dUTPase decreases the dUTP pool in the infected cell and thus prevents incorporation of dUTP into DNA during reverse transcription. *Pro* is a protease required for processing of some of the viral polypeptide precursors. *Pol* and *In* encode the reverse transcriptase (RT) and integrase genes respectively. The RT also has RNase H activity, which degrades the RNA strand in the newly synthesized RNA-DNA hybrid intermediates. The integrase catalyses the insertion of proviral DNA into the host DNA (reviewed by [11]). *Env* directs the synthesis of a 73 kDa glycosylated precursor protein, which is proteolytically cleaved by cellular furin proteases into the surface gp52 and transmembrane gp36 proteins. *Rem* (regulator of export) is a recently identified gene that encodes a protein that facilitates the export of unspliced viral mRNAs. *Rem*-like genes are also found in complex retrovirusses, like *Rev* in HIV [91] but not in simple retroviruses like Murine Leukemia Virus (MuLV). The *Sag* (superantigen) gene is transcribed from the U3 region in the 3′ LTR. The gene is unique for MMTV and essential for its normal live cycle. The existence of yet another accessory protein, Naf (negative-acting factor), encoded by the U3 open reading frame has been reported [130]. This protein may negatively affect expression of certain cellular genes, but it is yet poorly characterized and not further discussed here. The other genes will be discussed in more detail below.

3.1.4 Transcription and Translation of the Viral Genes

After insertion into the host DNA, various viral mRNAs are synthesized as depicted in Fig. 3.2. The main product is genome-size mRNA that starts at the U3/R junction and continues into the 3′ LTR where the R/U5 junction contains the polyadenylation site. Part of this RNA is incorporated into new virions, while the remaining part is used for translation of the *Gag, Dut/Pro, Pol/In* genes. These genes overlap in different reading frames leading to the Gag, Gag-Dut/Pro and Gag-Dut/Pro-Pol/In polypeptide precursors. The translation of the Gag-Dut/Pro precursor requires a ribosomal frame shift near the Gag translational stop codon, while expression of the Gag-Dut/Pro-Pol/In precursor polypeptide requires a double frame shift near the *Pro-Pol* junction [102]. The double frame shift needed to translate the latter precursor is rather unique for MMTV and is not required for most other retroviruses. These subsequent ribosomal frame shifts are carried out with low efficiency and provide a mechanism to reduce the ratio between the more 3′ located enzymatic active proteins that are required in minor amounts and the structural Gag proteins that are needed in much larger quantities. The 5′ end of *Dut* overlaps with the 3′ end of the *Gag* sequence, while both full length gene products are present in the *Gag-Dut/Pro* precursor (Fig. 3.2). Dut is a transframe protein that arises from a ribosomal frameshift at the *Gag-Dut/Pro* junction [72]. Also, the RT enzyme is a transframe protein containing at its N-terminus sequences from the C-terminus of *Pro* [29]. The various polypeptide precursors that arise from this complex mode of translation are subsequently proteolytically processed to release the active proteins.

The *Env* encoding mRNA also starts from the transcription start site at the U3/R boundary in the 5′ LTR but the *Gag-Dut/Pro-Pol/In* genes are spliced out using the 5′ splice donor (SD) and the first downstream splice acceptor (SA) site (Fig. 3.1). The *Sag* gene is translated from two spliced mRNAs. A minor Sag RNA species also starts from the standard transcription start site in the 5′ LTR, extends until the first SD site and continues from the second SA site into the U3 region of the 3′ LTR. The predominant *Sag* transcript starts from a transcription start site within the *Env* gene as depicted in Fig. 3.2 [106] As a consequence of the location of the transcription start sites of both Sag mRNAs, *Sag* can only be transcribed from the 3′ LTR in the provirus [17]. *Sag* transcription is driven by lymphoid lineage specific promoter elements within the *Env* and *Pol* genes [121] and thus expression is limited to lymphoid cells. *Rem* is produced from a double spliced mRNA, which partially overlaps in frame with the 5′ and 3′ ends of the *Env* genes, and the 5′ end of the *Sag* gene [91].

3.1.5 Virion Structure and Assembly

Dimeric viral genomic RNA is encapsulated by unprocessed Gag, Gag-Dut/Pro and Gag-Dut/Pro-Pol/In polypeptides to form a spherical core. This cytoplasmic RNA-protein core is termed intracytoplasmic A particle, which is an immature B-particle [136]. The viral core in turn is surrounded during the budding process by a plasma membrane domain enriched in Env glycoproteins (see below). Subsequently, the Gag containing precursors are proteolytically processed by Pro in at least six nonglycosylated proteins including the matrix, capsid and nucleocapsid proteins. The latter component directly interacts with the viral RNA. The Gag derived proteins, which are highly overrepresented relative to the proteins with enzymatic activity form the actual structural components of the core in which the enzymes are also included. The processing of the precursors into the respective functional proteins occurs in a very ordered fashion and is intimately linked to virion assembly, budding and maturation (reviewed by Vogt [157]). Processing of the Gag proteins results in a mature virion with an eccentric core (Fig. 3.1).

Env encoded peripheral membrane protein gp52 and transmembrane protein gp36 form a stable complex that is incorporated in the plasma membrane and may accumulate in certain membrane domains. Subsequently, the cytoplasmic domain of gp36 interacts with the N-terminally located future matrix protein in the unprocessed Gag polypetide of the intracytoplasmic A particle. Trimers of the gp52–gp36 heterodimer form the characteristic surface knobs of MMTV. Gp52 binds to the MMTV receptor on the cell surface during infection. A more detailed description of the viral gene expression and processing of these gene products is provided by Vogt et al. [157].

3.1.6 Infectious Life Cycle

The infectious life cycle of MMTV contains several features that are unique among retroviruses. Most remarkably, the virus exploits the cellular host immune system to ultimately infect mammary epithelial cells. MMTV virions ingested via the milk by new born mice travel through the intestinal wall and infect first dendritic cells (DCs) and subsequently B and T cells in the gut wall and Peyer's patches [21, 48]. Infection takes place upon binding of the envelope gp52/gp36 complex to the mouse transferrin receptor on the surface of the host cells [52, 129]. Subsequently, the viral lipid bilayer and host cell membrane fuse, delivering the core of the virus into the cytoplasm of the host cell. At this point, the RNA is reverse transcribed by the viral RT (included in the viral core), resulting in a double stranded DNA molecule with at each end identical long terminal repeat (LTR) regions as outlined above. This full length proviral DNA translocates to the nucleus where it integrates essentially randomly into the genome, catalyzed by the viral integrase. Next, cellular RNA polymerase II synthesizes the viral genomic RNA and various spliced messenger RNAs as discussed above and depicted in Fig. 3.2.

Unique for MMTV is the crucial role of the *Sag* gene during the early stages of infection of new born mice. In the infected professional antigen presenting cells, DCs and B cells, the *Sag* gene is expressed at the cell surface in association with the Major Histocompatibility Complex (MHC) class II proteins. The Sag-MHC complex strongly interacts with T cell receptors on CD4 positive T cells that contain particular Vβ chains and provokes a superantigen response that strongly stimulates the proliferation of the Sag-reactive T cells (reviewed by [1]). Cytokines released by the activated T cells in turn drive the expansion of the MMTV infected B cell population [49]. During this process, T cells are also infected and some of the infected B cells mature into long-lived memory cells. In this way, a reservoir of MMTV infected B and T cells is generated in suckling mice that can transport the virus to all parts of the body during a longer time period. This is an indispensable step for efficient infection of mammary epithelial cells since the virus cannot infect the resting, non-differentiated mammary epithelial cells until these cells start to proliferate at puberty. At that point, the infected circulating lymphocytes in the mammary gland will infect the expanding mammary epithelial cells [39]. As a consequence, MMTV cannot effectively infect DC or B cell deficient mice or mice that lack Sag specific T cells [21, 39]. Infection of mammary epithelial cells proceeds essentially the same as infection of cells in the hematopoietic lineage, except that *Sag* plays no role and expression of the viral proteins is largely hormonally regulated. As a consequence, after initial infection of mammary cells, the virus will remain largely repressed in the mammary gland until pregnancy and lactation when pregnancy hormones stimulate viral protein expression. This allows infectious virions to further spread through the mammary epithelium, thus allowing the virus to be subsequently secreted into the milk.

3.1.7 Tissue Specificity and Hormonal Regulation of MMTV Replication

The mammary gland is the predominant site of MMTV replication and consequently the main site of MMTV induced malignant transformation. Since the cellular receptor for MMTV binding and entry into the cell is the transferrin receptor, which is ubiquitously expressed on growing cells, the receptor is unlikely to determine the tropism of the virus. Indeed, in MMTV infected animals, viral proteins are also expressed at relatively low levels and in a strain dependent fashion in epithelial cells in a variety of other tissues such as salivary glands, kidneys, lungs, seminal vesicles, prostate, testes and in some lymphoid tissues [50, 58]. However, high expression levels are only found in the mammary epithelial cells. Current evidence indicates that mammary specific expression and replication of MMTV is primarily controlled at the transcriptional level. Composite enhancer elements that contribute to mammary cell-restricted expression have been mapped to the 5′ proximal end of the U3 region [41, 42, 168]. Negative regulatory elements (NREs) at the 3′ proximal end of the U3 region are involved in MMTV suppression in non-epithelial tissues including lymphoid cells [100, 128]. Both regions strongly determine the mammary gland specificity of MMTV. MMTV expression in the mammary gland is most abundant during pregnancy and lactation. Along with other factors, this is regulated by the homeodomain containing repressor protein (CDP) that binds to multiple sites in the NRE. The CDP protein is highly expressed in undifferentiated cells, but its expression is absent in differentiated cells, including those in the lactating mammary gland [170]. In addition, Stat5a and Stat5b, which are prolactin-inducible transcription factors, bind to the U3 domain and stimulate MMTV expression during lactation [117]. CDP and Stat5a/Stat5b act in concert to limit MMTV expression to the lactation period when the virus needs to be transmitted.

Viral gene expression is strongly dependent on steroid hormones [155] In particular, glucorticoids and progesterone strongly enhance the expression of viral proteins and replication of the virus in the mammary gland [15, 112, 124]. Without these hormones, the strength of the MMTV-LTR promoter is attenuated. As a consequence, MMTV infected ovariectomized mice rarely develop tumors. The steroid hormone dependent viral gene expression is caused by hormone response elements (HRE) located just upstream of the transcription start site within the U3 region of the LTR [15, 57, 131], and reviewed in by Gunzburg et al. [45]. The HRE contains at least four overlapping recognition sites to which both activated glucocorticoid and progestin receptors can bind [16]. The transcriptional repressor CDP also binds to sites overlapping with and adjacent to the steroid binding elements. This allows CDP to compete with steroid hormone binding, and thus further contributes to the pregnancy and lactation dependent expression [170]. In addition to the regulatory domains discussed above, several more regulatory domains within the LTR have been identified that contribute to viral protein expression regulation [100]. Moreover, the transcriptional activity of the genomic region where the proviral DNA integrates may also control MMTV expression and replication.

The pregnancy dependent regenerative cycles of the mammary gland is another factor that may influence MMTV replication, and consequently, MMTV induced tumorigenicity in the mammary gland as compared to other infected tissues such as the salivary gland. Although the mammary gland cellularity increases about 30-fold during each pregnancy, most cells die during involution of the gland after lactation ends. This suggests that mammary stem cells or lineage progenitors that persist after involution are the (oncogenic) targets of MMTV [73], since expansion of the mammary gland from these cells during subsequent pregnancies coincides with a strong increase of MMTV positive cells. However, the tumors induced by different virus strains can vary in hormone dependence (see below). Therefore, these MMTV strains may infect or replicate in different types of progenitor cells.

3.1.8 Exogenous MMTV

All mouse strains with a high-incidence of mammary tumors horizontally transmit infectious MMTV via milk to their offspring. These viruses are designated exogenous MMTV variants. Replication competent MMTV can also be vertically transmitted when a provirus stably integrates in cells of the germline. The first MMTV variants are exclusively transmitted via milk and were derived from mouse strains with a high mammary tumor incidence. These exogenous viruses are carried among others by the A, DBA, C3H, BR6, RIII and CzechII strains. The exogenous MMTV variants from these mouse strains are highly virulent, and thus most relevant for insertional mutagenesis. However, the exogenous MMTV variants are often lost in the present day germ-free maintained mouse strains as a result of foster-nursing on non-virus carriers. MMTV from the various mouse strains does not necessarily induce the same type of tumors. As will be discussed below, some MMTV strains induce hormone dependent tumors, while other virus strains mainly induce hormone independent tumors. Conversely, the same virus may induce different histological types of tumors in different mouse strains. For example, C3H-MMTV causes an approximately equal percentage of low grade acinar carcinomas (Dunn type A tumors) and low grade solid carcinomas (Dunn type B tumors) in C3H mice, while in BALB/c mice it induces almost exclusively carcinomas of the acinar type that have a relatively high propensity to metastasize to the lung [150].

3.1.9 Endogenous MMTV

Pups of high mammary tumor incidence strains (e.g., C3H) can be foster-nursed on mice without the milk factor. These "virus free" animals still develop mammary tumors, albeit with a longer latency and lower incidence, due to activation of endogenous proviral MMTV copies. Also, the tumors arising in low incidence strains (e.g., BALB/c) are also thought to be caused by activation of replication competent endogenous MMTV copies. In fact, the genome of almost all inbred mouse strains harbors multiple copies of endogenous MMTV proviruses, mostly replication defective, epigenetically silenced or inserted into transcriptionally inactive chromatin regions. The number of MMTV proviral copies varies per strain, ranging from none to as many as twenty independent proviral insertions. These proviral copies result from the occasional infection of germline cells and are present in all somatic cells of offspring that develop from these infected gametes. If not inactivated, the endogenous provirus will produce infectious virus. For example, the endogenous C3H virus responsible for the late mammary tumors referred to above, is encoded by a gene designated *Mtv1* on chromosome 7 [153]. GR mice carry five endogenous MMTV copies, of which one (*Mtv2*) is active and responsible for the high incidence of early mammary tumors in this strain. As a consequence, foster-nursing GR pups on a dame without milk-transmitted virus does not decrease the tumor incidence [154]. Moreover, the MMTV variant expressed by *Mtv2* can also be transmitted horizontally via the milk from a GR female to another strain, and subsequently propagated as an exogenous virus. In contrast to *Mtv1* and *Mtv2,* most other *Mtv* genes are partially or completely inactive [109]. A typical example of the latter gene is *Mtv8,* which is present in almost all inbred mouse strains. Other endogenous *Mtv* genes apparently have been incorporated in the germline only recently, and are thus unique to specific strains. The CzechII mouse strain is rather exceptional in that it harbors no endogenous MMTV copies [35], which has facilitated insertional mutagenesis screens in these mice using exogenous virus.

3.1.10 MMTV-Induced Tumors

MMTV-infected mice usually develop mammary tumors at a relatively young age and at high frequency. Tumor incidence and latency depends on the efficiency of viral infection and replication in the different mouse strains rather than on tumorigenesis itself [127]. Some strains, such as the C57Bl/6, are rather resistant to MMTV infection and tumor induction. Resistance can be due to the inability of MHC class II genes to present SAG, a crucial step required for infection. In some strains endogenous *Mtv* genes produce SAG protein that is autoreactive with the same Vβ T cell receptor chains as the infecting virus, and thus have deleted the responsive T cells that are required to establish a productive infection with exogenous MMTV [127].

In wild type mice, MMTV mainly induces two types of mammary tumors: low grade acinar carcinoma and low grade solid tumors (previously indicated as Dunn type A and B respectively) or mixtures of these histotypes, depending on the mouse strain [14]. These tumors are often considered as rather benign. However, the frequency of lung metastases in mammary tumor-bearing BALB/c mice that acquired the C3H MMTV by foster nursing (BALB/cfC3H) was reported to be 63 and 16% in BALB/cfRIII females [138]. In our own studies, we found that approximately one third of the BALB/cfC3H mammary tumor-bearing mice had developed lung metastases at the time the primary tumor had reached $\sim$1 cm^3 (J. Hilkens and M. Boer, unpublished results).

In some strains, MMTV initially induces premalignant mammary lesions and tumors that are pregnancy hormone dependent, while in other strains all MMTV induced palpable tumors are hormone independent. In the next paragraphs, we will briefly review the biology of MMTV induced tumors including hormone dependence.

3.1.11 MMTV Induced Hormone Dependent Premalignant Lesions

Mammary tumors in the mouse usually develop through premalignant hyperplastic lesions. The most frequent hyperplastic lesions in the mouse are hyperplastic alveolar nodules (HANs) and ductal hyperplasia's (designated plaques). HANs are commonly found in all high-incidence mammary tumor strains and are focal proliferations from lobuloalveolar mammary epithelium [23, 89]. HANs contain immortalized cells that can be propagated by serial transplantations in mammary fat pads cleared from normal epithelium [87]. At onset, HANs are hormone dependent but when palpable, HANs have usually become hormone independent. Plaques are ductal hyperplasia's that initially regress after each parturition but reappear at the same site in subsequent pregnancies. However, after several parities, regression of these nodules is less complete and they eventually progress to pregnancy independent tumors [139, 154]. Plaques are commonly present in high-incidence MMTV strains such as GR and RIII females. Note that the pregnancy hormone (estrogen and progesterone) dependent tumor growth discussed in this paragraph should not be confused with the pregnancy dependence of MMTV replication, which affects the frequency of mutagenic events and is mainly glucocorticoid dependent.

HANs and plaques are associated with MMTV infection and express large amounts of MMTV in high-incidence MMTV strains. The discovery that HANs and plaques show MMTV proviral insertions near Wnt and Fgf genes indicates that MMTV insertional mutagenesis in concert with hormonal stimulation give rise to these hyperplasia's. However, HANs can also be observed in mice treated with chemical carcinogens in low-incidence MMTV strains, indicating that development of mammary tumors from premalignant lesions is a general phenomenon (reviewed by [88]).

3.1.12 Hormone-Dependent Mammary Tumors and MMTV

Plaques in the GR mouse strain that develop into mammary tumors are initially estrogen dependent and do not grow in ovariectomized mice; instead they progress from a hormone-dependent growth phase to autonomous, hormone-independent tumors [151, 152]. Hormone-dependent mammary tumors in ovariectomized GR mice only develop upon treatment with progesterone and estrogen. Transplantation of hormone dependent GR tumors usually leads, after several passages, to the emergence of hormone independent tumor variants as a result of a clonal selection process [96, 135]. MMTV is essential for the development of these tumors as they do not arise in GR congenic mice that lack the active endogenous provirus *Mtv2*. Of note, *Wnt* and *Fgf* genes are already activated in GR tumors by insertional mutagenesis before they become hormone independent [92, 103]. This indicates that expression of these oncogenes does not cause hormone independence, but one or more additional insertional mutations events or other genetic or epigenetic changes confer the hormone independent growth.

Squartini et al. [139] demonstrated that the mammary tumors induced by different strains of exogenous virus progress differently. In their study, BALB/c mice, which have a low incidence of mammary tumors, show a high incidence of hormone-independent tumors when infected with C3H-MMTV but develop hormone-dependent tumors progressing to hormone-independence when infected with RIII-MMTV. These results show that development of hormone dependent tumors is virus strain dependent. The exact cause of this virus strain specific hormone dependence is currently unknown.

3.1.13 MMTV-Induced Tumors in other Tissues

As discussed above, expression and replication of MMTV is low in most non-mammary tissues because of its tissue and hormonal-specific regulation. This largely precludes proviral integrations and tumor induction in these tissues. However, MMTV infection of hematopoietic cells does occur, although the LTR enhancer/promoter is probably inefficient in these cells. Evidence that the regulatory sequences in the MMTV-LTR are crucial for its tissue specific oncogenic activity comes from relatively rare MMTV induced lymphomas that develop at a low frequency late in life in some MMTV infected mouse strains, such as the GR and DBA/2 [95]. The genome of these lymphomas contains many new MMTV proviral copies with consistent deletions and gains in the enhancer-like elements in the U3 region of the LTR that alter the behavior of the LTR enhancer/promoter in lymphocytes [5, 55, 97, 167]. Because these sequence changes in the MMTV-LTR require time to accumulate within an infected animal, lymphomas occur with a long latency and are only observed in males as the females succumb to mammary tumors at a younger age. The acquired mutations in the LTR of the lymphotropic MMTV more readily allow viral replication and proviral integration in lymphoid tissues, and this eventually causes lymphomas through insertional mutagenesis [7, 90, 167].

Based on these findings, Yanagawa et al. [167] and Bhadra et al. [7], were able to construct infectious B-type T cell tropic viruses that are poorly expressed in mammary epithelial cells and are not transmitted via the milk, but instead replicate only in lymphoid cells due to removal of the negative regulatory element and acquisition of T cell specific enhancer elements. Interestingly, the *Sag* gene is not required for the B-type T cell tropic virus, indicating that expression of *Sag* is only required to generate a B and T cell population that can infect mammary cells [105]. Insertional mutagenesis screens using B-type leukemogenic virus showed that this virus deregulates some of the same genes that are also frequently tagged by Moloney virus in lymphomas, such as *Myc* and *Rorc* [10, 119, 148].

MMTV-associated kidney carcinomas have also been observed in Balb/c mice infected with milk-borne C3H MMTV. The kidney tumors harbored newly integrated exogenous MMTV proviruses that showed striking alterations in the U3 region of the LTRs that might be responsible for the difference in tissue specificity of the virus [162]. Changes in the U3 region were also reported for newly acquired proviral copies in AtT-20 mouse pituitary tumor cells although it is unclear whether these copies are involved in tumorigenesis [118]. The reported tissue-specific alterations in the LTR predict that highly infectious MMTV versions with modified LTRs may be useful for insertional mutagenesis in non-mammary tissues.

3.2 Gene Discovery by MMTV Mediated Insertional Mutagenesis

3.2.1 Common Integration Sites

MMTV randomly integrates in the murine genome [32], but in contrast to some other retroviruses such as Moloney Murine Leukemia virus, MMTV does not show a preference for transcription start regions and CpG islands [166]. If the mutagenic event caused by an MMTV proviral insertion affects a gene that confers a selective growth advantage, it will expand the mutated cell population, which may lead to hyperplasia. In some cells, additional hits near other cancer genes may provide these cells with an additional growth advantage or other tumor related properties, and this will eventually lead to full-blown mammary tumors. Most tumors are oligoclonal for proviral insertions, which is caused by subsequent insertions in different cells of the early tumor that provide similar selective advantages during tumor progression. However, many if not most additional insertions are simply fortuitous, irrelevant insertions that are piggybacking on the growth promoting insertions when the tumor expands. When analyzing the pattern of MMTV insertion sites in a collection of mammary tumors, these background insertion events should show a random distribution pattern. By contrast, insertion events causally linked to tumor formation are predicted to impact a subset of genes (i.e. proto-oncogenes), and thus should appear as nonrandom clusters found near these genes. Therefore, only those insertions that are clustered more densely than would be expected by chance in a set of independent tumors are likely to be responsible for tumor induction. These insertions are considered to be present in common integration sites (CISs) and indicate the presence of a nearby oncogene.

Mikkers et al. [99] calculated the size of the genomic window in which two or more insertions can be expected to be nonrandom, and thus defined a CIS as a function of the number of analyzed tumors and detected insertions. However, modern high-throughput screens produce increasingly large data sets and consequently, the calculated window may become quite small and no longer biologically relevant since the viral enhancers may act over much larger distances than the calculated window. Similarly, two or more insertions within the same gene may have the same effect on gene function but may not be present within the same calculated window. Recently, a statistical framework based on Gaussian kernel convolution has been developed [22], that is capable of detecting CISs in a noisy and biased environment using a predefined significance level while controlling the probability of detecting false CISs. Gaussian kernel convolution (summing Gaussian shaped kernel functions placed at every RIS, akin to a sliding window), results in a smoothed density distribution of inserts over the entire genome. Depending on the width of the kernel function, closely spaced RISs give rise to high peaks, indicating CISs. The significance of the height of the peaks can be determined by comparison with the peak heights observed in randomly distributed RISs after appropriate multiple testing correction. By varying the width of the kernel function, CISs at any biologically relevant scale can be detected.

3.2.2 Modes by Which Insertional Mutagenesis Affects Cancer Genes

The LTRs of proviral MMTV contain promoter and enhancer sequences that not only control the transcriptional activation of the viral genes, but also can influence expression of adjacent host genes. There are three major modes by which a provirus can affect the expression of nearby host genes: (1) promoter insertion, (2) enhancement of a nearby promoter or (3) integration within a gene causing truncation of the transcript (see Fig. 3.3). For promoter insertion to take place, the proviral MMTV must integrate in close proximity of the 5′ end and in the same transcriptional orientation as the target gene. In this way, the promoter within the 3′ LTR of the provirus can replace the endogenous gene promoter. Transcription of an adjacent gene can also start from the 5′ LTR after deletion of the 3′ end of the provirus or by removal of most of the 3′ viral transcript by splicing.

Since transcriptional enhancers act in an orientation-independent manner, the strong enhancer elements within the U3 region of the 5′ LTR cannot only stimulate the viral promoter but also the promoter of an adjacent cellular gene. Transcriptional enhancement will occur when the provirus integrates either downstream of a cellular

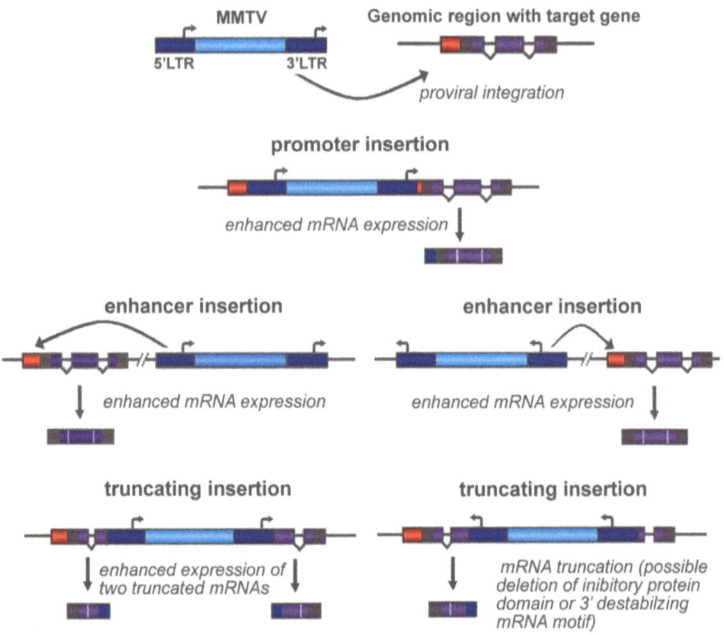

Fig. 3.3 Mechanisms of retroviral insertional mutagenesis. The most frequent mechanisms of retroviral insertional mutagenesis are depicted. Promoter insertion: the provirus inserts upstream of the target gene in the same transcriptional orientation and the viral promoter replaces the endogenous promoter of the target gene, which may lead to enhanced expression of the target gene. Variants of this mechanism may lead to alternative spliced forms of the mRNA by removal of one or more of the 5′ exons (not shown). Enhancer insertion: insertions 5′ in the reverse orientation or 3′ in the same orientation relative to the target gene, which may lead to activation of the endogenous promoter and subsequently to enhanced mRNA expression. Truncating insertion: Insertion in an exon or intron within the gene; these insertions may lead to inactive proteins, to enhanced expression of 5′ or 3′ truncated mRNAs resulting in N- or C-terminal truncated proteins that may be constitutively active due to removal of inhibitory domains, or to mRNA stabilization by removal of destabilizing motifs in the 3′ mRNA. The novel 5′ truncated messenger will terminate on a cryptic poly-adenylation site within the proviral LTR (not indicted). The *light* and *dark blue* bars represents MMTV proviral DNA enclosed by two LTRs (*dark blue*). The *purple bars* represent the open reading frame of the target gene; the promoter and untranslated regions of the target gene are depicted by *red* and *black bars* respectively. The figure is based on [62, 145]

gene in the same transcriptional orientation or upstream of a gene in the reverse orientation. Transcriptional enhancement is the most frequent mode of gene activation by retroviral insertion, as enhancement allows more flexibility with respect to the orientation of the virus and the distance to the target gene. The range by which enhancement can occur is illustrated by our finding that the normally silent Dpp10 gene is activated by proviral insertion at a distance of no less than 700 kb [144].

Intragenic insertions that lead to truncation of the gene can have different outcomes. For instance, integration within the 3′ end of a gene can result in enhanced expression of the gene by removal of destabilizing signals within the 3′ untranslated region of its mRNA. MMTV insertion in the same transcriptional orientation within the coding region of a gene may give rise to truncated transcripts that encode N- or C-terminal protein truncations. The enhancer in the 5′ LTR can stimulate transcription of a truncated 5′ transcript of the target gene, while the promoter in the 3′ LTR can drive transcription of a truncated 3′ transcript if there is a splice acceptor site in 3′ end of the gene (Fig. 3.3). Whether any of these truncated proteins are active obviously depends on the target gene and the exact insertion site. Intragenic proviral integration may also lead to an inactive gene that may promote tumorigenesis if the affected target gene is a tumor suppressor. However, a proviral insertion only affects a single allele of a gene. Thus an inactivating insertion in the second allele required to effectuate the oncogenic potential of a tumor suppressor gene is statistically very rare. However, an inactivating insertion in a single allele of a haploinsufficient tumor suppressor gene may lead to a selective advantage. Also, subsequent inactivation of the second tumor suppressor allele by loss of heterozygosity or promoter methylation could lead to clonal outgrowth. However, a selective advantage due to provirally-induced haploinsufficiency is apparently rather rare. For instance, haploinsufficiency of *Pten* is known to cause mammary tumorigenesis in mice and collaborates with *Wnt1* activation [78], one of the most frequent MMTV targets. Nevertheless, we did not find any proviral MMTV insertion in the *Pten* locus among approximately 2000 proviral MMTV insertions in mouse mammary tumors (Vendel-Zwaagstra A., Theodorou V. and Hilkens J., unpublished). Also, other known human or mouse tumor suppressor genes such as *Rb1* were not associated with common insertion sites in our screen. The genomic regions containing these genes could be poor targets for proviral integration, as these genes were also rarely found near 10 000 CIS in a MuLV screen of 510 lymphomas [148]. Interestingly, in their analysis almost half the MuLV insertions were intragenic, of which several occurred in genes that are known tumor suppressors such as *Ikaros*, *Nf1* and *E2f2*; the latter being a known haploinsufficient gene.

The above outline describes the most common mechanisms of by which genes can be activated but other more exotic mechanisms have also been described [62, 145, 147].

As discussed, the MMTV promoter activity is greatly enhanced by steroid hormones (e.g. glucocorticoids, progestins) and is weak without hormones. The question arises whether this will affect expression of the targeted oncogene. Sonnenberg *et al.* [137], showed that activation of *Fgf2* by proviral MMTV insertion (RIII-derived) in the opposite transcriptional orientation was hormone-independent, while MMTV expression in the same cells remained hormone-dependent. Grimm and Nordeen [41] investigated this further using luciferase reporter assays and confirmed that activation of the promoter of at least some cellular MMTV target genes by the MMTV-LTR enhancer (C3H derived) is relatively independent of steroid hormones.

3.3 Gene Families and Pathways Frequently Targeted by MMTV

The first oncogenes identified using MMTV-induced insertional mutagenesis were discovered in conventional MMTV positive inbred mouse strains such as the C3H, RIII, BR6, GR and CzechII strains. The first identified MMTV common integration site was located adjacent to the *int-1* gene, which later was identified as the ortholog of the *wingless* gene in *Drosophila* [123], and was subsequently designated *Wnt1*. A second common MMTV-insertion site, designated *int-2* [113], was found to encode a member of the fibroblast growth factor family (*Fgf3*) [24]. A frequent MMTV target in CzechII mice, *int-3* [35] was later recognized as a member of the Notch family, *Notch4* [34, 125]. These first three MMTV targets are members of three different gene families critical for normal patterning during embryogenesis. Presently, three additional members of the Wnt family, *Wnt3* [126], *Wnt3a* 144] and *Wnt-10b* [77], and three additional Fgf-family members, *Fgf4* [114], *Fgf8* [132], *Fgf10* [143], and one additional Notch-family member, *Notch1* [25] have been associated with MMTV common insertion sites.

The early studies showed that usually only a few members within a gene family are MMTV targets. Also in our high-throughput IM study [144], in mammary tumors from MMTV infected BALB/c mice, we only rarely found that other members of the three above mentioned gene families were tagged, suggesting that specific gene family members are oncogenic in the mammary gland. We will tentatively refer to the MMTV targets within a gene family as the "oncogenic" family members. However, it remains possible that tagging of other members of these gene families does not confer a selective advantage due to the absence of the appropriate receptors in mammary epithelial cells (in case of secreted factors), constrains on the interaction between the target gene promoter and MMTV enhancer or the presence of some of these gene family members in transcriptionally inactive DNA regions. The former option appears unlikely for Fgfs as some non-targeted Fgfs use the same Fgf receptors as targeted Fgfs. The latter option seems also unlikely as the targeted Wnt members are present in evolutionary conserved, closely spaced pairs of which only one is targeted by MMTV.

Genes that are transcriptionally activated by proviral insertion are usually silent or expressed at very low levels in normal postnatal mammary epithelium. Therefore, the distinction between oncogenic and non-oncogenic family members is further supported by the fact that only non-oncogenic *Wnt* and *Fgf*-genes are expressed in postnatal normal mammary epithelial cells (e.g. *Wnt4*, *Wnt6*), and thus are unlikely to be oncogenic unless their normal activity is kept in check by a strictly temporal regulation (e.g. only during pregnancy). In that case, however, it can easily be envisaged that MMTV insertion would interfere with such temporal regulation. Interestingly, *Wnt10b* was reported to be an MMTV target [77], despite that it is expressed in normal mammary epithelium [75, 77]. However, in our own IM studies, we noticed that many insertions predicted to activate *Wnt10b* also, and occasionally even exclusively, activate the nearby *Wnt1* gene. Conversely, the majority of the tumors expressed *Wnt10b* without insertion in or nearby the *Wnt10b*/*Wnt1* locus [144]. Therefore, *Wnt10b* is probably not a bona fide MMTV target and is not considered to be an strongly oncogenic Wnt, despite the fact that high overexpression in transgenic mice leads to mammary tumors [75].

Our high-throughput IM study and subsequent expression analysis (M. Kimm, M. Boer and J. Hilkens, unpublished) showed that only rarely is more than one oncogenic *Wnt* or *Fgf* gene expressed in individual tumors. Clearly, activation of a single member of each oncogenic family is sufficient to fully activate the respective pathways and activation of additional genes in the same pathway provide no further selective advantage in most cases, although some exceptions to this rule will be discussed bellow. In nearly all MMTV induced BALB/c mammary tumors an oncogenic *Wnt* gene and an oncogenic *Fgf* are co-targeted, suggesting interdependence of both gene families in tumorigenesis. A preference for specific combinations of Wnt and Fgf family members was not observed despite some reports to the contrary [64].

There is some evidence that MMTV insertions within multiple genes in the same pathway provide a selective advantage when they occur within a single tumor. The Fgf-receptor1 (*Fgfr1*) and *Fgfr2* genes are occasionally tagged in the same tumors in which an Fgf ligand is also tagged. Apparently, the Fgf receptor is not expressed at sufficiently high levels, and therefore enhanced expression as a result of MMTV tagging can provide a further selective advantage. Whether the tumors without MMTV-induced Fgfr expression use other ways to upregulate the appropriate receptor has not been investigated. This finding shows that cooperativity between MMTV target genes can occur within the same pathway under certain circumstances.

Another example of genes collaborating within the same pathway is provided by the R-spondin genes. Two genes of the R-spondin gene family, *Rspo2* and *Rspo3*, are frequently associated with CISs [37, 81, 144]. The R-spondin gene family comprises four members that are characterized by a thrombospondin type 1 domain and two furin domains. Normal mammary epithelial cells transduced with cDNA encoding *Rspo3* are tumorigenic in vivo, validating this MMTV target as a novel proto-oncogene [144]. In particular, *Rspo2* and *Rspo3* amplify *Wnt1* and *Wnt3a* signaling, while R-spondins alone have only a minor effect on the Wnt-pathway [65, 67, 107]. Rspo's bind to the Wnt co-receptor LRP6 [8, 161],and antagonize DKK1 activity by interfering with DKK mediated LRP6 and Kremen association, leading to down regulation of the Frizzled-LRP receptor complex [68, 108]. In other words, activation of an Rspo gene may be required to prevent DKK-mediated inhibition of the Wnt pathway. Indeed, MMTV insertions in *Rspo2* and *Rspo3* are frequently found together with a Wnt gene insertion in the same tumor (Vendel-Zwaagstra and Hilkens, unpublished) and represent another example how cooperativity between genes in the same pathway can provide a selective advantage to the tumor cell.

Notch4 is targeted with high frequency in CzechII mammary tumors but only at low frequently in MMTV induced BALB/c tumors. However, activation of the Notch pathway may be underestimated, as other genes in the Notch signaling cascade are occasionally tagged in the Balb/c tumors [35, 144]. Fgf and Wnt genes were usually activated in the tumors in which Notch is provirally tagged. This suggests that these pathways are collaborating, although Notch pathway activation does not seem to be a prerequisite for Wnt/Fgf induced mammary tumorigenesis in BALB/c mice.

3.3.1 Infrequently Targeted Genes and Pathways

Recently, we have published a list of 33 candidate genes in MMTV-induced tumors found in 160 wild type BALB/c mice [144]. Insertional mutagenesis in additional mouse mammary tumor models extended this list to more than 40 CISs (A. Vendel-Zwaagstra, M. Boer and J. Hilkens, unpublished). In addition to the frequently targeted genes acting in the Wnt, Fgf and Notch pathways that are discussed above, we also found a large series of infrequently tagged genes not acting in one of these pathways ([144], and A. Vendel-Zwaagstra, M. Boer and J. Hilkens, unpublished). Among these are genes acting in a variety of pathways such as Igf2/insulin pathway including *Igf2* and *Irs4*, members of the Odz, Robo and glypican gene families, *Eras*, *Id2*, *Map3K8, Astn2, Sfmbt2, Dpp10, Zfp521* and others. The function of most of these genes in mammary tumorigenesis has not been elucidated.

Some of the infrequently targeted genes may in fact activate the same pathway. For instance, among the infrequently tagged genes are three genes that may activate the PI3Kinase pathway: *Irs4,* a member of the Insulin Receptor Substrate family [76], *Igf2* a secreted growth factor that among others activates the PI3K pathway and *Eras* which directly activates the PI3K pathway [141]. An expression analysis shows that these genes are expressed in 15–20% of the mammary tumors depending on the genetic background, although this analysis does not necessarily represent the tagging frequency but may better indicate the importance of these genes. Several genes acting in this signaling cascade are well established cancer genes, like *PIK3CA*, encoding the catalytic moiety of the PI3-kinase and the tumor suppressor gene *PTEN* (reviewed in [26]). Loss of PTEN expression or activating *PIK3CA* mutations occur in more than 50% of the human breast cancers and Pten loss was also shown to accelerate mouse mammary tumorigenesis [78]. Of note is that *Irs4* and *Igf2* may also be involved in other oncogenic pathways, like the MAPK/ERK pathway.

Theodorou et al. [144] showed that "single" insertion sites (insertion sites present in only a single tumor) can still be informative because some genes may be tagged infrequently as they activate less dominant pathways, or the same pathway may be activated by multiple genes from the same gene family decreasing the probability to be present in a CIS. Indeed, analysis of the complete dataset of MMTV retroviral insertion sites by using various bioinformatics tools and public databases revealed that certain gene families, protein domains, and pathways in addition to the ones associated with CISs are overrepresented among genes associated with single insertion sites. Gene family analysis of 160 MMTV-induced BALB/c tumors revealed that among others Ras, cyclin, cadherins and T-box gene families are overrepresented. Protein domain analysis of the genes tagged by single insertions showed that genes encoding proteins with a thrombospondin type 1 (TSP-1), phosphokinase, or cation ATPase domain were also overrepresented and likely important in carcinogenesis. Proteins with a TSP-1 domain also include *Rspo2* and *Rspo3*, mentioned above, and several *Adamts* family members some of which also have been implicated in cancer [101]. Based on these results, we can also consider genes present in "common gene families" as candidate cancer genes.

3.4 Comparison of MMTV Targets in Mammary Tumors and MuLV Targets in Lymphomas

Theodorou et al. [144] investigated the overlap between the entire sets of genes targeted by MMTV and MuLV (deposited in the Mouse Retrovirus Tagged Cancer Gene Database, RTCGD, http://rtcgd.abcc.ncifcrf.gov/mm8/index.html) and whether genes targeted by both viruses are acting in the same or in different pathways. The comparison showed that MMTV and MuLV target genes and signaling pathways only partially overlap. In lymphomas MuLV frequently tags genes involved in the development of the hematopoietic and immune system, whereas MMTV preferentially tags genes related to mammary development. This could either be due to the nature of the different retroviruses or to the tissues that are infected by the viruses. Comparing the MuLV targets in brain tumors present in the RTCGD with the MuLV targets in lymphomas suggest that tumor formation is mainly dictated by the nature of the tissue rather than the nature of the virus.

3.5 Core and Sporadic Activated Cancer Pathways

In most mammary tumors, several CISs are present. Some of these CISs occur with very high frequency while others are only infrequently found. Callahan and Smith [13] designated the former as core CISs. The core CISs activate mainly the Wnt, Fgf and Notch pathways in a large proportion of MMTV induced mammary tumors although the frequency may vary somewhat (for Wnt, Fgf) or significantly (for Notch) in different mouse strains and may also depend on the MMTV variant used to induce the tumors. Consistent with Callahan and Smith, we will designate these pathways as the core oncogenic pathways in MMTV-induced tumors in contrast to the infrequently activated pathways, which we will designate as sporadic pathways. The essential role of the core pathways in mouse mammary tumorigenesis is clearly illustrated in wild-type Balb/c tumors in which Wnts and Fgfs are activated in over 90% of the cases. Although both pathways act synergistically in bitransgenic mice, together these pathways cannot drive tumorigenesis as *Wnt1/Fgf3* bitransgenic mice development tumors with latencies between 3 and 10 months indicating additional genetic or epigenetic events are required [74, 53]. Therefore, it is not surprising that in almost all tumors we found MMTV proviral insertions in addition to those activating core pathways. However, these insertions tag a variety of genes with low frequency, of which some are in CISs. These insertions activate genes that participate in various pathways that probably collaborate with one or more core pathways. Interestingly, both *Fgf3* and *Wnt1* transgenic and bitransgenic mice develop hyperplastic mammary glands, suggesting that the additional insertions may cause the malignant transformation. However, these hits may only contribute to tumor progression [13]. The clonality of the sporadically activated genes in the tumor may provide some clues in this respect.

In the next paragraphs, we will briefly discuss the core pathways activated by MMTV proviral insertion in mammary epithelial tissues.

3.5.1 Wnt Signaling

Wnt signaling plays a key role in many aspects of embryogenesis and throughout adult life by regulating morphology, proliferation, motility, cell fate and stem cell maintenance. Somatic mutations in components of the Wnt pathway are also frequently seen in a variety of human cancers, although only infrequently in human breast cancer. However, nuclear β-catenin (a read out for activation of the canonical Wnt-pathway) has been observed in 60% of tumors, indicating the importance of this pathway also for human breast cancer [79]. Wnts are lipid modified secreted molecules. Nineteen highly conserved Wnt genes have been identified in mice and humans, but only some Wnts have been shown to be involved in mouse tumorigenesis. Also, the transforming capacity of the various Wnt genes in vitro varies greatly [164, 134]. In addition to the MMTV-tagged Wnts, also Wnt2 and Wnt7A can transform C57MG cells in vitro.

Wnt signaling induces stabilization of β-catenin, which is subsequently translocated to the nucleus where it acts as cofactor for TCF/LEF transcription factors. TCF/LEF transcription factors then regulate genes involved in cell proliferation such as *cyclinD1* and *c-Myc*. This is considered the canonical Wnt-pathway, recently reviewed by Clevers [19]. In addition, several non-canonical pathways have been described including the planar cell polarity and cell migration pathways through activation of Rho family GTPases, Jnk and the Ca^{2+} pathway through phospholipase C activation.

Wnts act through Frizzled receptors, of which no less than ten have been identified. However, Wnts can also bind to other membrane molecules acting as Frizzled coreceptors or as independent receptors—including *Lrp5*, *Lrp6*, *Ror2*, *Ryk* and proteoglycan cofactors such as syndecans and glypicans [66]. Recent evidence shows that the pathway that is engaged by Wnts depends on the specific receptor-coreceptor complex present at the cell surface and the cellular context [149]. The canonical pathway is only activated by a Frizzled-Lrp complex, whereas the non-canonical Wnt signaling pathway may involve Frizzled receptors, but in association with different coreceptors [149].

Only the canonical Wnt pathway seems to be involved in mouse mammary tumorigenesis as ectopic expression of stabilized β-catenin in the mammary gland is sufficient to cause mammary tumors that are indistinguishable from Wnt induced tumors [94]. While the canonical pathway is involved in initiation, the non-canonical pathways may only contribute to tumor cell migration and invasiveness [60]. In our MMTV insertional mutagenesis screen we only find genes acting in the canonical pathway such as the Rspo genes but no genes known to act in one of the non-canonical pathways. This again indicates that the canonical pathway is one of the dominant oncogenic pathways in mouse mammary tumorigenesis.

Mice transgenic for *Wnt1* driven by the MMTV-LTR display accelerated proliferation of the mammary ductal and alveolar epithelium resulting in lobuloalveolar hyperplasia throughout the mammary glands and the development of focal mammary tumors with a median latency of approximately 5 months [146]. The development of generalized mammary hyperplasia and subsequent solitary mammary tumors with a relatively long latency suggest that Wnt gene activation is an early event in mammary tumorigenesis. Subsequent collaborating genetic events are required for malignant tumor formation. However, continued Wnt stimulation remains required for tumor maintenance and growth [44]. Involvement of Wnts in tumor initiation is in keeping with the observation that Wnt gene activation in MMTV induced mammary tumors is an early event [146].

3.5.2 *Fgf Signaling*

In mammals, the fibroblast growth factor gene family comprises of 22 members. Nearly all Fgf genes are expressed at some point during embryonic development and are essential for organogenesis and limb development. Fgf signaling results in a variety of cellular responses, like proliferation, migration and survival (for review see [30, 43, 116]). Fgfs are secreted molecules that signal in an autocrine as well as paracrine fashion through cell-surface tyrosine kinase receptors encoded by four genes (*Fgfr1-4*). Alternative splicing of the *Fgfr1, 2* and *3* transcripts results in at least 7 variants, which bind the various Fgfs with different affinities. The various receptors show different biological activities which adds to the complexity of Fgf signaling [116]. Activation of the FGFRs also involves polysulphated proteoglycans like heparin or heparan (HSPG) that bind Fgfs and are required as accessory receptors further enhancing the complexity of Fgf-signaling. Upon ligand binding, FGFRs dimerize and autophosphorylate specific tyrosine residues in their cytoplasmic domains. The phosphorylated domains form docking sites for molecules involved a various signaling pathways of which the Ras/MAPK, PI3K/Akt and PLCγ pathways are the most prominent [30]. In vitro, FGFs are involved in cell proliferation, migration, differentiation and survival. In vivo, some FGFs have strong angiogenic activity while others are involved in tissue remodeling and wound repair. Fgf overexpression in tumor cells could act as an autocrine growth factor and/or as an angiogenic factor. Interestingly, the Fgf family members with the strongest angiogenic capacity, such as *Fgf1* and *Fgf2*, are never targeted by MMTV, suggesting that the main contribution of FGFs in mouse mammary tumorigenesis might be stimulation of cell proliferation and/or survival.

Mammary glands of MMTV-*Fgf3* transgenic mice exhibit extensive epithelial proliferation and develop ductal hyperplasia, which only in some transgenic strains gives rise to tumors relatively late in life [74, 77, 104]. As discussed above, Fgf and Wnt genes are usually co-targeted by MMTV, suggesting that both gene families closely collaborate. Wnt and Fgf genes are already co-mutated in mammary hyperplasia, which shows that activation of both genes are early events in tumorigenesis

[103, 115]. Indeed, Wnt1 and Fgf3 transgenic mice develop generalized diffuse mammary hyperplasia confirming that Wnts and Fgfs are involved in early stages of carcinogenesis [104, 146].

The compound MMTV-Wnt1/MMTV-*Fgf3* transgenic mice also developed tumors with a shorter latency than transgenic mice expressing only a single transgene, proving that Wnt and Fgf signaling collaborate in mammary tumorigenesis. However, additional genetic and/or epigenetic events are required for tumorigenesis [53, 74]. Also, our insertional mutagenesis studies show that together with Wnt and Fgf genes a variety of additional genes are activated that are likely to collaborate with these genes.

3.5.3 Notch Signaling

Notch signaling determines cell fate decisions such as differentiation, proliferation and apoptosis. The Notch family includes four transmembrane receptors, *Notch1-4*, and five ligands, *Jag1*, *Jag2*, *Dll1*, *Dll3*, and *Dll4*. In addition, numerous modulators of the pathway have been described. Notch receptors are synthesized as precursor proteins, which upon cleavage form a non-covalently bound heterodimer at the cell surface. Notch ligands are present at the surface of adjacent cells, which means that Notch signaling is only limited to neighboring cells. Upon ligand binding to the receptor, the Notch intracellular domain (N-ICD) is proteolytically released and is translocated to the nucleus. Subsequently, N-ICD binds to the transcription factor CSL, which releases the CSL-bound repressor and activates transcription of target genes (reviewed in [71]). Hes and Hey family members are among the N-ICD/CSL target genes, but it has also been shown that the *Myc* oncogene along with the cell cycle regulators *Cdkn1a* ($p21^{waf1}$) and *Ccnd1* are targets of Notch signaling [70, 98].

Proviral insertion in *Notch4* is always intragenic in such a way that the inserted provirus leads to an N-terminal truncated Notch protein that only contains the transmembrane and intracellular domains (in essence the N-ICD). The inserted MMTV promoter constitutively drives the expression of this truncated Notch form. This gain-of-function mutation mimics ligand-activated Notch4. As a result, MMTV proviral insertion enables activation of Notch target genes in a ligand-independent manner. Also MuLV targeting of Notch genes in lymphomas leads to a gain-of-function by the same mechanism. While *Notch2* and *Notch3* are not targeted by MMTV, similar intragenic proviral insertions as in *Notch4* were also found in *Notch1* in MMTV induced tumors from *ErbB2* transgenic mice [25]. The contribution of Notch signaling to cancer has been linked to the suppression of apoptosis through activation of AKT [93]. Transgenic mice expressing the truncated *Notch4* genomic fragment encoding the ICD under control of the MMTV-LTR show impaired mammary gland differentiation and develop poorly differentiated adenocarcinomas of the mammary gland [36, 61]. Interestingly, similar transgenic mice expressing human *Notch1*-ICD first show lactation-dependent tumors that regress after weaning, but eventually develop hormone-independent non-regressing adenocarcinomas [70].

3.6 MMTV Targets and Mammary Stem Cells

Notch and Wnt signaling have been implicated in mammary stem cell self-renewal [3, 27]. In MMTV-*Wnt1* transgenic mice, the number of stem cells is expanded over 6-fold suggesting that *Wnt*-signaling participates in controlling the self-renewal of mammary stem cells [133]. Notch and Wnt involvement in mammary stem self-renewal is reminiscent of its function in hematopoietic stem cells. Also, the Fgfs (e.g. *Fgf4*) have been implicated in stem cell maintenance. FGFs are commonly used to expand tissue specific stem cells and embryonic stem cells (reviewed in [40]). The Hedgehog signaling pathway has also been implicated in mammary stem cell self-renewal [80]. However, so far only one component, *Ptchd3,* was found to be infrequently tagged by MMTV, suggesting that this pathway plays a minor role in mouse mammary tumorigenesis. It is often speculated that deregulation of stem cell self-renewal is one of the key events involved in carcinogenesis. The fact that three MMTV induced core oncogenic pathways also act in stem cell self-renewal supports this notion.

3.7 Relevance of Insertional Mutagenesis for Human Breast Cancer

At first sight, many frequently activated genes in MMTV-induced mouse mammary tumors from wild-type mice are distinct from those commonly mutated in human breast cancers. This may not be surprising as over 60% of human breast cancers are hormone-dependent, while most mouse mammary tumors are hormone-independents. Also the histopathology is different. Most human breast tumors are invasive ductal and lobular carcinomas, whereas most mammary tumors in the mouse are low grade and rather benign. However, the histology of many mammary tumors that arise in genetically modified mouse strains is often much more human like. One of the most "humanized" mammary tumors arises in the *ErbB2* transgenic mouse strains. These tumors not only histological mimic human Neu/Her2 tumors, but also show similarity in gene expression profiles [2]. Therefore, it can be expected that genes activated by insertional mutagenesis in these models are more relevant for human breast cancer than genes activated in wild-type mice.

It is also becoming increasingly clear that activation of the oncogenic pathways is of more universal importance than the individually activated genes. For instance, several of the genes initially discovered by insertional mutagenesis, such as *Wnt1*, are only rarely expressed in human breast cancers. However, the canonical Wnt-pathway is activated in a large proportion of human breast cancers as shown by stabilization of β-catenin and upregulation *Axin2* and *Lef1* [4, 54, 79]. Moreover, activation of the Wnt-pathway in human primary mammary epithelial cells leads to transformation in vitro and tumor formation upon orthotopic transplantation into immunocompromised mice [4]. Several lines of evidence indicate that the Wnt pathway in human breast cancers is frequently activated by epigenetic silencing of inhibitory genes in the pathway (reviewed in [69]). In contrast, in colon cancer the Wnt pathway is most often activated by mutations in the *APC* gene, preventing degradation of β-catenin. These and similar observations illustrate that activation of the same oncogenic pathways in different species, or even different tissues, can occur through different mechanisms. Therefore, the pathways affected by MMTV-induced mutations in mouse models may be more indicative for human breast cancer than the MMTV target genes themselves.

In some pathways, however, the same genes are affected in mouse and human cancer. For instance, the human orthologs of the MMTV targets *Fgf8, Fgf10, Fgfr1* and *Fgfr2* are overexpressed and *Fgf3* and *Fgfr2* are frequently amplified in human breast cancer [85, 143, 169]. The latter gene is also associated with a high risk of local-regional recurrence in node-negative cancers [33]. Genome-wide association studies using single nucleotide polymorphisms identified *FGFR2* as a breast cancer susceptibility gene [28, 56]. In conclusion, genes in the Fgf-pathway are oncogenic targets in mouse and human breast cancer, further emphasizing the Fgf-pathway as a key oncogenic pathway.

In the Notch pathway, some components such as Notch genes themselves are MMTV targets and are also associated with several human cancers. Most notably, activating mutations in *NOTCH1* are present in over 50% of the T-cell acute lymphoblastic leukemias [163]. In a small study of twenty human breast carcinomas, accumulation of the Notch-ICD was found in all samples [140]. Also loss of *NUMB*, a negative regulator of Notch signalling, and expression of known downstream target genes was observed indicating that aberrantly activated Notch signalling is a frequent event in breast carcinomas [140]. Moreover, patients with tumors expressing high levels of *JAG1* or *NOTCH1* had a significantly poorer overall survival compared with patients expressing low levels of these genes [120]. Notch ligands, however, are only rarely tagged by MMTV.

Taken together, all three core oncogenic pathways that are activated by MMTV insertional mutagenesis in mouse mammary tumors are also relevant for human breast cancer. However, it seems unlikely that all three core pathways are relevant for all different breast cancer subtypes. For instance, in our own study we rarely find Wnt or Fgf tagging in mammary tumors derived from MMTV infected *ErbB2* transgenic mice.

Igf2 and *Irs4* are both acting in the insulin-like signalling pathway and are present among the infrequently MMTV tagged genes ([144], Zwaagstra, Boer and Hilkens, unpublished). Igf2 is a paternal imprinted gene in humans and mice. Loss of imprinting of *Igf2* has been observed in a significant proportion of human breast cancers [86]. Also, enhanced levels of circulating *Igf1* has been implicated breast cancer [47]. Although both *Igf2* and *Igf1* activate the insulin-like growth factor-1 receptor, *Igf1* is not an MMTV target. Possibly, *IGF2* is more critical for tumorigenesis as *IGF1* is secreted by the normal mammary stroma, while the stroma surrounding the malignant mammary epithelium switches expression to *IGF2* [111]. *IRS4* was found to be mutated in eleven breast cancer samples tested by Wood et al. [165], suggesting that it is also relevant in human breast cancer.

Of the 33 genes associated with novel MMTV CISs reported by Theodorou et al. [144], seven sporadic activated genes (*ASTN2*, *FGFR2*, *JMJD1C*, *DPP10*, *NOTCH4*, *ODZ1*, *PROS1*, *RREB1*) were found to be mutated in eleven human breast tumor samples tested in a screen for mutated genes [165]. In addition, we found in our own gene expression analysis using microarray data from 295 breast carcinomas that expression of *ASTN2, CENTG2, EGR3, FGFR2, GSE1, JMJD1C, IGF2 LAMB1, PDGFRB, PROS1, RREB1* was deregulated in more than 5% of the cancers. The large overlap between the affected genes in both studies indicates the significance for human breast cancer of the sporadic tagged genes in the mouse models.

3.8 Concluding Remarks

Recent years have shown that insertional mutagenesis screens, including the screens for MMTV proviral insertions, lead to the discovery of a wealth of novel candidate cancer genes. Many of the target genes act in known human oncogenic pathways indicating the validity of insertional mutagenesis screens for human breast cancer. However, the number of more sporadically activated genes that are validated as genuine cancer genes is rather limited. This validation, although urgently required, may not be simple. Most sporadic target genes are likely to act as cancer gene only in concert with one or more specific gene mutations. In the next years to come, more functional analysis of individual genes and gene sets should be performed to fully benefit from the present screens. Finding the pathways in which some of the uncharacterized proviral MMTV target genes act may greatly profit from gene expression analyses in the same tumors. Combining these technologies, however, has not yet been explored on a large scale, but has a great potential. In fact, it may be most useful as a rapid method to determine whether a candidate gene participates in a novel oncogenic pathway or simply activates one of the known core pathways. At present, insertional mutagenesis, as well as tumor genome sequencing, have revealed an enormous complexity and diversity of genes involved in mammary tumorigenesis. One of the questions that must be solved is whether this complexity can be reduced when these genes are assigned to an oncogenic pathway.

Presently most techniques to identify retroviral or transposon insertions are based on conventional sequencing of individually amplified sequences adjacent to the insert. In addition to the collection of the tumor samples, this is one of the limiting steps in the identification of novel CIS. The new "next generation" sequencing methods allow the parallel sequencing of all insertions in a single tumor with high sensitivity and potentially can give a good indication of the clonality of a specific insert. The clonality is a measure to distinguish genes that act early and late in tumorigenesis, which is yet difficult to determine. Tumors may be more addicted to early events than to the relatively late events that are more likely involved in tumor progression and properties related to invasion and metastasis.

Insertional mutagenesis identifies genes and signaling pathways causally related to mammary tumors in the mouse. One drawback of insertional mutagenesis is that it can only be performed in mouse models, although this may be largely solved using improved genetic modified mouse models for human cancer. As we have shown above, a large proportion of MMTV target genes are also deregulated or mutated in human breast cancer, while other genes act in oncogenic pathways known to be involved in human breast cancer. So this drawback of insertional mutagenesis may be not so serious after all. CGH and mutation analyses of the entire genomes of human tumors may largely benefit from the insertional mutagenesis studies. In fact, databases containing the genomic landscapes of mammary tumors and the retroviral targets may be cross compared to exclude genes with passenger mutations and vice versa to confirm the relevance of MMTV targets in mouse mammary tumors for human breast cancer. Similarly, the relevant cancer genes in the amplified genomic regions defined by CGH studies of human breast cancers may be identified using insertional mutagenesis data.

Presently, it is unlikely that we have obtained the complete picture of all relevant pathways in mammary carcinogenesis. Novel candidate cancer genes may be found in various genetically modified mouse models. It can be envisaged that the use of transposon tagging identifies additional candidate mammary cancer genes for instance when the transposons is provided with different (non-LTR) promoters or if the transposon is mobilized in different progenitor cells in the mammary gland. Such screens may aid to elaborate the interplay between different oncogenic pathways, which is required to develop more efficient and specific anti-cancer therapies and to develop new prognostic and predictive indicators.

References

1. Acha-Orbea, H., & MacDonald, H. R. (1995). Superantigens of mouse mammary tumor virus. *Annual Reviews of Immunology*, *13*, 459–486.
2. Andrechek, E. R., Laing, M. A., Girgis-Gabardo, A. A., Siegel, P. M., Cardiff, R. D., & Muller, W. J. (2003). Gene expression profiling of neu-induced mammary tumors from transgenic mice reveals genetic and morphological similarities to ErbB2-expressing human breast cancers. *Cancer Research*, *63*, 4920–4926.
3. Androutsellis-Theotokis, A., Leker, R. R., Soldner, F., Hoeppner, D. J., Ravin, R., Poser, S. W., et al. (2006). Notch signalling regulates stem cell numbers in vitro and in vivo. *Nature*, *442*, 823–826.
4. Ayyanan, A., Civenni, G., Ciarloni, L., Morel, C., Mueller, N., Lefort, K., et al. (2006). Increased Wnt signaling triggers oncogenic conversion of human breast epithelial cells by a Notch-dependent mechanism. *Proceedings of the National Academy of Sciences USA*, *103*, 3799–3804.
5. Ball, J. K., Diggelmann, H., Dekaban, G. A., Grossi, G. F., Semmler, R., Waight, P. A., et al. (1988). Alterations in the U3 region of the long terminal repeat of an infectious thymotropic type B retrovirus. *Journal of Virology*, *62*, 2985–2993.

6. Bannert, N., & Kurth, R. (2004). Retroelements and the human genome: New perspectives on an old relation. *Proceedings of the National Academy of Sciences USA*, *101*(Suppl 2), 14572–14579.
7. Bhadra, S., Lozano, M. M., & Dudley, J. P. (2005). Conversion of mouse mammary tumor virus to a lymphomagenic virus. *Journal of Virology*, *79*, 12592–12596.
8. Binnerts, M., Kim, K., Bright, J., Patel, S., Tran, K., Zhou, M., et al. (2007). R-Spondin1 regulates Wnt signaling by inhibiting internalization of LRP6. *Proceedings of the National Academy of Sciences USA*, *104*, 14700–14705.
9. Bittner, J. J. (1936). Some possible effects of nursing on the mammary gland tumor incidence in mice. *Science*, *84*, 162.
10. Broussard, D. R., Lozano, M. M., & Dudley, J. P. (2004). Rorgamma (Rorc) is a common integration site in type B leukemogenic virus-induced T-cell lymphomas. *Journal of Virology*, *78*, 4943–4946.
11. Brown, P. O. (1997). Integration. In J. M. Coffin, S. H. Hughes, & H. E. Varmus (Eds.), *Retroviruses* (pp. 161–203). Cold Spring Harbor Laboratory Press (NY), USA.
12. Callahan, R., & Smith, G. H. (2000). MMTV-induced mammary tumorigenesis: Gene discovery, progression to malignancy and cellular pathways. *Oncogene*, *19*, 992–1001.
13. Callahan, R., & Smith, G. H. (2008). Common integration sites for MMTV in viral induced mouse mammary tumors. *Journal of Mammary Gland Biology and Neoplasia*, *13*, 309–321.
14. Cardiff, R. D., Anver, M. R., Gusterson, B. A., Hennighausen, L., Jensen, R. A., Merino, M. J., et al. (2000). The mammary pathology of genetically engineered mice: The consensus report and recommendations from the annapolis meeting. *Oncogene*, *19*, 968–988.
15. Cato, A. C., Miksicek, R., Schutz, G., Arnemann, J., & Beato, M. (1986). The hormone regulatory element of mouse mammary tumour virus mediates progesterone induction. *EMBO Journal*, *5*, 2237–2240.
16. Chalepakis, G., Arnemann, J., Slater, E., Bruller, H. J., Gross, B., & Beato, M. (1988). Differential gene activation by glucocorticoids and progestins through the hormone regulatory element of mouse mammary tumor virus. *Cell*, *53*, 371–382.
17. Choi, Y., Kappler, J. W., & Marrack, P. (1991). A superantigen encoded in the open reading frame of the 3' long terminal repeat of mouse mammary tumour virus. *Nature*, *350*, 203–207.
18. Chu, E. Y., Hens, J., Andl, T., Kairo, A., Yamaguchi, T. P., Brisken, C., et al. (2004). Canonical WNT signaling promotes mammary placode development and is essential for initiation of mammary gland morphogenesis. *Development*, *131*, 4819–4829.
19. Clevers, H. (2006). Wnt/beta-catenin signaling in development and disease. *Cell*, *127*, 469–480.
20. Collier, L. S., Carlson, C. M., Ravimohan, S., Dupuy, A. J., & Largaespada, D. A. (2005). Cancer gene discovery in solid tumours using transposon-based somatic mutagenesis in the mouse. *Nature*, *436*, 272–276.
21. Courreges, M. C., Burzyn, D., Nepomnaschy, I., Piazzon, I., & Ross, S. R. (2007). Critical role of dendritic cells in mouse mammary tumor virus in vivo infection. *Journal of Virology*, *81*, 3769–3777.
22. de Ridder, J., Uren, A., Kool, J., Reinders, M., & Wessels, L. (2006). Detecting statistically significant common insertion sites in retroviral insertional mutagenesis screens. *PLoS Computational Biology*, *2*, e166.
23. DeOme, K. B., Faulkin, L. J., Jr., Bern, H. A., & Blair, P. B. (1959). Development of mammary tumors from hyperplastic alveolar nodules transplanted into gland-free mammary fat pads of female C3H mice. *Cancer Research*, *19*, 515–520.
24. Dickson, C., & Peters, G. (1987). Potential oncogene product related to growth factors. *Nature*, *326*, 833.
25. Dievart, A., Beaulieu, N., & Jolicoeur, P. (1999). Involvement of Notch1 in the development of mouse mammary tumors. *Oncogene*, *18*, 5973–5981.
26. Dillon, R. L., White, D. E., & Muller, W. J. (2007). The phosphatidyl inositol 3-kinase signaling network: Implications for human breast cancer. *Oncogene*, *26*, 1338–1345.

27. Dontu, G., Jackson, K. W., McNicholas, E., Kawamura, M. J., Abdallah, W. M., & Wicha, M. S. (2004). Role of Notch signaling in cell-fate determination of human mammary stem/progenitor cells. *Breast Cancer Research*, *6*, R605–R615.
28. Easton, D. F., Pooley, K. A., Dunning, A. M., Pharoah, P. D., Thompson, D., Ballinger, D. G., et al. (2007). Genome-wide association study identifies novel breast cancer susceptibility loci. *Nature*, *447*, 1087–1093.
29. Entin-Meer, M., Avidan, O., & Hizi, A. (2003). The mature reverse transcriptase molecules in virions of mouse mammary tumor virus possess protease-derived sequences. *Virology*, *310*, 157–162.
30. Eswarakumar, V. P., Lax, I., & Schlessinger, J. (2005). Cellular signaling by fibroblast growth factor receptors. *Cytokine and Growth Factor Reviews*, *16*, 139–149.
31. Fantozzi, A., & Christofori, G. (2006). Mouse models of breast cancer metastasis. *Breast Cancer Research*, *8*, 212.
32. Faschinger, A., Rouault, F., Sollner, J., Lukas, A., Salmons, B., Gunzburg, W. H., et al. (2008). Mouse mammary tumor virus integration site selection in human and mouse genomes. *Journal of Virology*, *82*, 1360–1367.
33. Fioravanti, L., Cappelletti, V., Coradini, D., Miodini, P., Borsani, G., Daidone, M. G., et al. (1997). Int-2 oncogene amplification and prognosis in node-negative breast carcinoma. *International Journal of Cancer*, *74*, 620–624.
34. Gallahan, D., & Callahan, R. (1997). The mouse mammary tumor associated gene INT3 is a unique member of the NOTCH gene family (NOTCH4). *Oncogene*, *14*, 1883–1890.
35. Gallahan, D., & Callahan, R. (1987). Mammary tumorigenesis in feral mice: Identification of a new int locus in mouse mammary tumor virus (Czech II)-induced mammary tumors. *Journal of Virology*, *61*, 66–74.
36. Gallahan, D., Jhappan, C., Robinson, G., Hennighausen, L., Sharp, R., Kordon, E., et al. (1996). Expression of a truncated Int3 gene in developing secretory mammary epithelium specifically retards lobular differentiation resulting in tumorigenesis. *Cancer Research*, *56*, 1775–1785.
37. Gattelli, A., Zimberlin, M. N., Meiss, R. P., Castilla, L. H., & Kordon, E. C. (2006). Selection of early-occurring mutations dictates hormone-independent progression in mouse mammary tumor lines. *Journal of Virology*, *80*, 11409–11415.
38. Goedert, J. J., Rabkin, C. S., & Ross, S. R. (2006). Prevalence of serologic reactivity against four strains of mouse mammary tumour virus among US women with breast cancer. *British Journal of Cancer*, *94*, 548–551.
39. Golovkina, T. V., Dudley, J. P., & Ross, S. R. (1998). B and T cells are required for mouse mammary tumor virus spread within the mammary gland. *Journal of Immunology*, *161*, 2375–2382.
40. Gotoh, N. (2009). Control of stemness by fibroblast growth factor signaling in stem cells and cancer stem cells. *Current Stem Cell Research and Therapy*, *4*, 9–15.
41. Grimm, S. L., & Nordeen, S. K. (1998). Mouse mammary tumor virus sequences responsible for activating cellular oncogenes. *Journal of Virology*, *72*, 9428–9435.
42. Grimm, S. L., & Nordeen, S. K. (1999). A composite enhancer element directing tissue-specific expression of mouse mammary tumor virus requires both ubiquitous and tissue-restricted factors. *Journal of Biological Chemistry*, *274*, 12790–12796.
43. Grose, R., & Dickson, C. (2005). Fibroblast growth factor signaling in tumorigenesis. *Cytokine Growth Factor Reviews*, *16*, 179–186.
44. Gunther, E. J., Moody, S. E., Belka, G. K., Hahn, K. T., Innocent, N., Dugan, K. D., et al. (2003). Impact of p53 loss on reversal and recurrence of conditional Wnt-induced tumorigenesis. *Genes and Development*, *17*, 488–501.
45. Gunzburg, W. H., & Salmons, B. (1992). Factors controlling the expression of mouse mammary tumour virus. *Biochemical Journal*, *283*(Pt 3), 625–632.
46. Hanahan, D., & Weinberg, R. A. (2000). The hallmarks of cancer. *Cell*, *100*, 57–70.

47. Hankinson, S. E., Willett, W. C., Colditz, G. A., Hunter, D. J., Michaud, D. S., Deroo, B., et al. (1998). Circulating concentrations of insulin-like growth factor-I and risk of breast cancer. *Lancet, 351*, 1393–1396.
48. Held, W., Shakhov, A. N., Izui, S., Waanders, G. A., Scarpellino, L., MacDonald, H. R., et al. (1993a). Superantigen-reactive CD4+ T cells are required to stimulate B cells after infection with mouse mammary tumor virus. *Journal of Experimental Medicine, 177*, 359–366.
49. Held, W., Waanders, G. A., Shakhov, A. N., Scarpellino, L., Acha-Orbea, H., & MacDonald, H. R. (1993b). Superantigen-induced immune stimulation amplifies mouse mammary tumor virus infection and allows virus transmission. *Cell, 74*, 529–540.
50. Henrard, D., & Ross, S. R. (1988). Endogenous mouse mammary tumor virus is expressed in several organs in addition to the lactating mammary gland. *Journal of Virology, 62*, 3046–3049.
51. Hilkens, J. (2006). Recent translational research: Oncogene discovery by insertional mutagenesis gets a new boost. *Breast Cancer Research, 8*, 102.
52. Hilkens, J., van der, Z. B., Buijs, F., Kroezen, V., Bleumink, N., & Hilgers, J. (1983). Identification of a cellular receptor for mouse mammary tumor virus and mapping of its gene to chromosome 16. *Journal of Virology, 45*, 140–147.
53. Hollmann, C. A., Kittrell, F. S., Medina, D., & Butel, J. S. (2001). Wnt-1 and int-2 mammary oncogene effects on the beta-catenin pathway in immortalized mouse mammary epithelial cells are not sufficient for tumorigenesis. *Oncogene, 20*, 7645–7657.
54. Howe, L. R., & Brown, A. M. (2004). Wnt signaling and breast cancer. *Cancer Biology and Therapy., 3*, 36–41.
55. Hsu, C. L., Fabritius, C., & Dudley, J. (1988). Mouse mammary tumor virus proviruses in T-cell lymphomas lack a negative regulatory element in the long terminal repeat. *Journal of Virology, 62*, 4644–4652.
56. Hunter, D. J., Kraft, P., Jacobs, K. B., Cox, D. G., Yeager, M., Hankinson, S. E., et al. (2007). A genome-wide association study identifies alleles in FGFR2 associated with risk of sporadic postmenopausal breast cancer. *Nature Genetics, 39*, 870–874.
57. Hynes, N., van Ooyen, A. J., Kennedy, N., Herrlich, P., Ponta, H., & Groner, B. (1983). Subfragments of the large terminal repeat cause glucocorticoid-responsive expression of mouse mammary tumor virus and of an adjacent gene. *Proceedings of the National Academy of Sciences of the USA, 80*, 3637–3641.
58. Imai, S., Morimoto, J., Tsubura, Y., Iwai, Y., Okumoto, M., Takamori, Y., et al. (1983). Tissue and organ distribution of mammary tumor virus antigens in low and high mammary cancer strain mice. *European Journal of Cancer and Clinical Oncology, 19*, 1011–1019.
59. Jemal, A., Siegel, R., Ward, E., Hao, Y., Xu, J., Murray, T., et al. (2008). Cancer statistics. *CA Cancer Journal of Clinicians, 58*, 71–96.
60. Jessen, J. R. (2009). Noncanonical Wnt signaling in tumor progression and metastasis. *Zebrafish, 6*, 21–28.
61. Jhappan, C., Gallahan, D., Stahle, C., Chu, E., Smith, G. H., Merlino, G., et al. (1992). Expression of an activated Notch-related int-3 transgene interferes with cell differentiation and induces neoplastic transformation in mammary and salivary glands. *Genes and Development, 6*, 345–355.
62. Jonkers, J., & Berns, A. (1996). Retroviral insertional mutagenesis as a strategy to identify cancer genes. *Biochimica et Biophysica Acta, 1287*, 29–57.
63. Jonkers, J., & Berns, A. (2002). Conditional mouse models of sporadic cancer. *Nature Reviews Cancer, 2*, 251–265.
64. Kapoun, A. M., & Shackleford, G. M. (1997). Preferential activation of Fgf8 by proviral insertion in mammary tumors of Wnt1 transgenic mice. *Oncogene, 14*, 2985–2989.
65. Kazanskaya, O., Glinka, A., Barco Barrantes, I., Stannek, P., Niehrs, C., & Wu, W. (2004). R-Spondin2 is a secreted activator of Wnt/beta-catenin signaling and is required for Xenopus myogenesis. *Developmental Cell, 7*, 525–534.

66. Kikuchi, A., Yamamoto, H., & Sato, A. (2009). Selective activation mechanisms of Wnt signaling pathways. *Trends in Cell Biology, 19*, 119–129.
67. Kim, K. A., Kakitani, M., Zhao, J., Oshima, T., Tang, T., Binnerts, M., et al. (2005). Mitogenic influence of human R-spondinl on the intestinal epithelium. *Science, 309*, 1256–1259.
68. Kim, K. A., Wagle, M., Tran, K., Zhan, X., Dixon, M. A., Liu, S., et al. (2008). R-Spondin family members regulate the Wnt pathway by a common mechanism. *Molecular Biology of the Cell, 19*, 2588–2596.
69. Klarmann, G. J., Decker, A., & Farrar, W. L. (2008). Epigenetic gene silencing in the Wnt pathway in breast cancer. *Epigenetics, 3*, 59–63.
70. Klinakis, A., Szabolcs, M., Politi, K., Kiaris, H., Artavanis-Tsakonas, S., & Efstratiadis, A. (2006). Myc is a Notch1 transcriptional target and a requisite for Notch1-induced mammary tumorigenesis in mice. *Proceedings of the National Academy Of Sciences of theUSA, 103*, 9262–9267.
71. Kopan, R., & Ilagan, M. X. (2009). The canonical Notch signaling pathway: Unfolding the activation mechanism. *Cell, 137*, 216–233.
72. Koppe, B., Menendez-Arias, L., & Oroszlan, S. (1994). Expression and purification of the mouse mammary tumor virus gag-pro transframe protein p30 and characterization of its dUTPase activity. *Journal of Virology, 68*, 2313–2319.
73. Kordon, E. C., & Smith, G. H. (1998). An entire functional mammary gland may comprise the progeny from a single cell. *Development, 125*, 1921–1930.
74. Kwan, H., Pecenka, V., Tsukamoto, A., Parslow, T. G., Guzman, R., Lin, T. P., et al. (1992). Transgenes expressing the Wnt-1 and int-2 proto-oncogenes cooperate during mammary carcinogenesis in doubly transgenic mice. *Molecular and Cellular Biology, 12*, 147–154.
75. Lane, T. F., & Leder, P. (1997). Wnt-10b directs hypermorphic development and transformation in mammary glands of male and female mice. *Oncogene, 15*, 2133–2144.
76. Lavan, B. E., Fantin, V. R., Chang, E. T., Lane, W. S., Keller, S. R., & Lienhard, G. E. (1997). A novel 160-kDa phosphotyrosine protein in insulin-treated embryonic kidney cells is a new member of the insulin receptor substrate family. *Journal of Biological Chemistry, 272*, 21403–21407.
77. Lee, F. S., Lane, T. F., Kuo, A., Shackleford, G. M., & Leder, P. (1995). Insertional mutagenesis identifies a member of the Wnt gene family as a candidate oncogene in the mammary epithelium of int-2/Fgf-3 transgenic mice. *Proceedings of the National Academy of Sciences of the USA, 92*, 2268–2272.
78. Li, Y., Podsypanina, K., Liu, X., Crane, A., Tan, L. K., Parsons, R., et al. (2001). Deficiency of Pten accelerates mammary oncogenesis in MMTV-Wnt-1 transgenic mice. *BMC Molecular Biology, 2*, 2.
79. Lin, S. Y., Xia, W., Wang, J. C., Kwong, K. Y., Spohn, B., Wen, Y., et al. (2000). Beta-catenin, a novel prognostic marker for breast cancer: its roles in cyclin D1 expression and cancer progression. *Proceedings of the National Academy of Sciences of the USA, 97*, 4262–4266.
80. Liu, S., Dontu, G., Mantle, I. D., Patel, S., Ahn, N. S., Jackson, K. W., et al. (2006). Hedgehog signaling and Bmi-1 regulate self-renewal of normal and malignant human mammary stem cells. *Cancer Research, 66*, 6063–6071.
81. Lowther, W., Wiley, K., Smith, G. H., & Callahan, R. (2005). A new common integration site, Int7, for the mouse mammary tumor virus in mouse mammary tumors identifies a gene whose product has furin-like and thrombospondin-like sequences. *Journal of Virology, 79*, 10093–10096.
82. MacArthur, C. A., Shankar, D. B., & Shackleford, G. M. (1995). Fgf-8, activated by proviral insertion, cooperates with the Wnt-1 transgene in murine mammary tumorigenesis. *Journal of Virology, 69*, 2501–2507.
83. Mant, C., Hodgson, S., Hobday, R., D'Arrigo, C., & Cason, J. (2004). A viral aetiology for breast cancer: Time to re-examine the postulate. *Intervirology, 47*, 2–13.

84. Marchetti, A., Buttitta, F., Miyazaki, S., Gallahan, D., Smith, G. H., & Callahan, R. (1995). Int-6, a highly conserved, widely expressed gene, is mutated by mouse mammary tumor virus in mammary preneoplasia. *Journal of Virology*, *69*, 1932–1938.
85. Marsh, S. K., Bansal, G. S., Zammit, C., Barnard, R., Coope, R., Roberts-Clarke, D., et al. (1999). Increased expression of fibroblast growth factor 8 in human breast cancer. *Oncogene*, *18*, 1053–1060.
86. McCann, A. H., Miller, N., O'Meara, A., Pedersen, I., Keogh, K., Gorey, T., et al. (1996). Biallelic expression of the IGF2 gene in human breast disease. *Human Molecular Genetics*, *5*, 1123–1127.
87. Medina, D. (1996). The mammary gland: A unique organ for the study of development and tumorigenesis. *Journal of Mammary Gland Biology and Neoplasia*, *1*, 5–19.
88. Medina, D. (2000). The preneoplastic phenotype in murine mammary tumorigenesis. *Journal of Mammary Gland Biology and Neoplasia*, *5*, 393–407.
89. Medina, D. (2008). Premalignant and malignant mammary lesions induced by MMTV and chemical carcinogens. *Journal of Mammary Gland Biology and Neoplasia*, *13*, 271–277.
90. Mertz, J. A., Mustafa, F., Meyers, S., & Dudley, J. P. (2001). Type B leukemogenic virus has a T-cell-specific enhancer that binds AML-1. *Journal of Virology*, *75*, 2174–2184.
91. Mertz, J. A., Simper, M. S., Lozano, M. M., Payne, S. M., & Dudley, J. P. (2005). Mouse mammary tumor virus encodes a self-regulatory RNA export protein and is a complex retrovirus. *Journal of Virology*, *79*, 14737–14747.
92. Mester, J., Wagenaar, E., Sluyser, M., & Nusse, R. (1987). Activation of int-1 and int-2 mammary oncogenes in hormone-dependent and -independent mammary tumors of GR mice. *Journal of Virology*, *61*, 1073–1078.
93. Meurette, O., Stylianou, S., Rock, R., Collu, G. M., Gilmore, A. P., & Brennan, K. (2009). Notch activation induces Akt signaling via an autocrine loop to prevent apoptosis in breast epithelial cells. *Cancer Research*, *69*, 5015–5022.
94. Michaelson, J. S., & Leder, P. (2001). beta-catenin is a downstream effector of Wnt-mediated tumorigenesis in the mammary gland. *Oncogene*, *20*, 5093–5099.
95. Michalides, R., Wagenaar, E., Hilkens, J., Hilgers, J., Groner, B., & Hynes, N. E. (1982a). Acquisition of proviral DNA of mouse mammary tumor virus in thymic leukemia cells from GR mice. *Journal of Virology*, *43*, 819–829.
96. Michalides, R., Wagenaar, E., & Sluyser, M. (1982b). Mammary tumor virus DNA as a marker for genotypic variance within hormone-responsive GR mouse mammary tumors. *Cancer Research*, *42*, 1154–1158.
97. Michalides, R., Wagenaar, E., & Weijers, P. (1985). Rearrangements in the long terminal repeat of extra mouse mammary tumor proviruses in T-cell leukemias of mouse strain GR result in a novel enhancer-like structure. *Molecular and Cellular Biology*, *5*, 823–830.
98. Miele, L. (2006). Notch signaling. *Clinical Cancer Research*, *12*, 1074–1079.
99. Mikkers, H., Allen, J., Knipscheer, P., Romeijn, L., Hart, A., Vink, E., et al. (2002). High-throughput retroviral tagging to identify components of specific signaling pathways in cancer. *Nature Genetics*, *32*, 153–159.
100. Mink, S., Ponta, H., & Cato, A. C. (1990). The long terminal repeat region of the mouse mammary tumour virus contains multiple regulatory elements. *Nucleic Acids Research*, *18*, 2017–2024.
101. Mochizuki, S., & Okada, Y. (2007). ADAMs in cancer cell proliferation and progression. *Cancer Science*, *98*, 621–628.
102. Moore, R., Dixon, M., Smith, R., Peters, G., & Dickson, C. (1987). Complete nucleotide sequence of a milk-transmitted mouse mammary tumor virus: Two frameshift suppression events are required for translation of gag and pol. *Journal of Virology*, *61*, 480–490.
103. Morris, D. W., Barry, P. A., Bradshaw, H. D., Jr., & Cardiff, R. D. (1990). Insertion mutation of the int-1 and int-2 loci by mouse mammary tumor virus in premalignant and malignant neoplasms from the GR mouse strain. *Journal of Virology*, *64*, 1794–1802.

104. Muller, W. J., Lee, F. S., Dickson, C., Peters, G., Pattengale, P., & Leder, P. (1990). The int-2 gene product acts as an epithelial growth factor in transgenic mice. *EMBO Journal, 9*, 907–913.
105. Mustafa, F., Bhadra, S., Johnston, D., Lozano, M., & Dudley, J. P. (2003). The type B leukemogenic virus truncated superantigen is dispensable for T-cell lymphomagenesis. *Journal of Virology, 77*, 3866–3870.
106. Mustafa, F., Lozano, M., & Dudley, J. P. (2000). C3H mouse mammary tumor virus superantigen function requires a splice donor site in the envelope gene. *Journal of Virology, 74*, 9431–9440.
107. Nam, J. S., Turcotte, T. J., Smith, P. F., Choi, S., & Yoon, J. K. (2006). Mouse cristin/R-spondin family proteins are novel ligands for the Frizzled 8 and LRP6 receptors and activate beta-catenin-dependent gene expression. *Journal of Biological Chemistry, 281*, 13247–13257.
108. Niehrs, C. (2006). Function and biological roles of the dickkopf family of Wnt modulators. *Oncogene, 25*, 7469–7481.
109. Nusse, R., de Moes, J., Hilkens, J., & van Nie, R. (1980). Localization of a gene for expression of mouse mammary tumor virus antigens in the GR/Mtv-2- mouse strain. *Journal of Experimental Medicine, 152*, 712–719.
110. Nusse, R., & Varmus, H. E. (1982). Many tumors induced by the mouse mammary tumor virus contain a provirus integrated in the same region of the host genome. *Cell, 31*, 99–109.
111. Paik, S. (1992). Expression of IGF-I and IGF-II mRNA in breast tissue. *Breast Cancer Research and Treatment, 22*, 31–38.
112. Parks, W. P., Ransom, J. C., Young, H. A., & Scolnick, E. M. (1975). Mammary tumor virus induction by glucocorticoids. Characterization of specific transcriptional regulation. *Journal of Biological Chemistry, 250*, 3330–3336.
113. Peters, G., Brookes, S., Smith, R., & Dickson, C. (1983). Tumorigenesis by mouse mammary tumor virus: Evidence for a common region for provirus integration in mammary tumors. *Cell, 33*, 369–377.
114. Peters, G., Brookes, S., Smith, R., Placzek, M., & Dickson, C. (1989). The mouse homolog of the hst/k-FGF gene is adjacent to int-2 and is activated by proviral insertion in some virally induced mammary tumors. *Proceedings of the National Academy of Science of the USA, 86*, 5678–5682.
115. Peters, G., Lee, A. E., & Dickson, C. (1984). Activation of cellular gene by mouse mammary tumour virus may occur early in mammary tumour development. *Nature, 309*, 273–275.
116. Powers, C. J., McLeskey, S. W., & Wellstein, A. (2000). Fibroblast growth factors, their receptors and signaling. *Endocrine Related Cancer, 7*, 165–197.
117. Qin, W., Golovkina, T. V., Peng, T., Nepomnaschy, I., Buggiano, V., Piazzon, I., et al. (1999). Mammary gland expression of mouse mammary tumor virus is regulated by a novel element in the long terminal repeat. *Journal of Virology, 73*, 368–376.
118. Racevskis, J., & Beyer, H. (1989). Amplification of mouse mammary tumor virus genomes in non-mammary tumor cells. *Journal of Virology, 63*, 456–459.
119. Rajan, L., Broussard, D., Lozano, M., Lee, C. G., Kozak, C. A., & Dudley, J. P. (2000). The c-myc locus is a common integration site in type B retrovirus-induced T-cell lymphomas. *Journal of Virology, 74*, 2466–2471.
120. Reedijk, M., Odorcic, S., Chang, L., Zhang, H., Miller, N., McCready, D. R., et al. (2005). High-level coexpression of JAG1 and NOTCH1 is observed in human breast cancer and is associated with poor overall survival. *Cancer Research, 65*, 8530–8537.
121. Reuss, F. U., & Coffin, J. M. (1998). Mouse mammary tumor virus superantigen expression in B cells is regulated by a central enhancer within the pol gene. *Journal of virology, 72*, 6073–6082.
122. Richert, M. M., Schwertfeger, K. L., Ryder, J. W., & Anderson, S. M. (2000). An atlas of mouse mammary gland development. *Journal of Mammary Gland Biology and Neoplasia, 5*, 227–241.

123. Rijsewijk, F., Schuermann, M., Wagenaar, E., Parren, P., Weigel, D., & Nusse, R. (1987). The Drosophila homolog of the mouse mammary oncogene int-1 is identical to the segment polarity gene wingless. *Cell, 50*, 649–657.
124. Ringold, G. M., Yamamoto, K. R., Tomkins, G. M., Bishop, M., & Varmus, H. E. (1975). Dexamethasone-mediated induction of mouse mammary tumor virus RNA: A system for studying glucocorticoid action. *Cell, 6*, 299–305.
125. Robbins, J., Blondel, B. J., Gallahan, D., & Callahan, R. (1992). Mouse mammary tumor gene int-3: a member of the notch gene family transforms mammary epithelial cells. *Journal of Virology, 66*, 2594–2599.
126. Roelink, H., Wagenaar, E., Lopes, dS., & Nusse, R. (1990). Wnt-3, a gene activated by proviral insertion in mouse mammary tumors, is homologous to int-1/Wnt-1 and is normally expressed in mouse embryos and adult brain. *Proceedings of theNational Academy of Sciences of theUSA, 87*, 4519–4523.
127. Ross, S. R. (2008). MMTV infectious cycle and the contribution of virus-encoded proteins to transformation of mammary tissue. *Journal of Mammary Gland Biology and Neoplasia, 13*, 299–307.
128. Ross, S. R., Hsu, C. L. L., Choi, Y., Mok, E., & Dudley, J. P. (1990). Negative regulation in correct tissue-specific expression of mouse mammary tumor virus in transgenic MICE. *Molecular and Cellular Biology (November), 10*, 5822–5829.
129. Ross, S. R., Schofield, J. J., Farr, C. J., & Bucan, M. (2002). Mouse transferrin receptor 1 is the cell entry receptor for mouse mammary tumor virus. *Proceedings of the National Academy of Sciences of the USA, 99*, 12386–12390.
130. Salmons, B., Erfle, V., Brem, G., & Gunzburg, W. H. (1990). naf, a trans-regulating negative-acting factor encoded within the mouse mammary tumor virus open reading frame region. *Journal of virology, 64*, 6355–6359.
131. Scheidereit, C., Geisse, S., Westphal, H. M., & Beato, M. (1983). The glucocorticoid receptor binds to defined nucleotide sequences near the promoter of mouse mammary tumour virus. *Nature, 304*, 749–752.
132. Shackleford, G. M., MacArthur, C. A., Kwan, H. C., & Varmus, H. E. (1993). Mouse mammary tumor virus infection accelerates mammary carcinogenesis in Wnt-1 transgenic mice by insertional activation of int-2/Fgf-3 and hst/Fgf-4. *Proceedings of the National Academy of Sciences of the USA, 90*, 740–744.
133. Shackleton, M., Vaillant, F., Simpson, K. J., Stingl, J., Smyth, G. K., Asselin-Labat, M. L., et al. (2006). Generation of a functional mammary gland from a single stem cell. *Nature, 439*, 84–88.
134. Shimizu, H., Julius, M. A., Giarre, M., Zheng, Z., Brown, A. M., & Kitajewski, J. (1997) Transformation by Wnt family proteins correlates with regulation of beta-catenin. *Cell Growth and Differentiation, 8*, 1349–1358.
135. Sluyser, M., & van Nie, R. (1974). Estrogen receptor content and hormone-responsive growth of mouse mammary tumors. *Cancer Research, 34*, 3253–3257.
136. Smith, G. H. (1978). Evidence for a precursor-product relationship between intracytoplasmic a particles and mouse mammary tumour virus cores. *Journal of General Virology, 41*, 193–200.
137. Sonnenberg, A., van Balen, P., Hilgers, J., Schuuring, E., & Nusse, R. (1987). Oncogene expression during progression of mouse mammary tumor cells; activity of a proviral enhancer and the resulting expression of int-2 is influenced by the state of differentiation. *EMBO Journal, 6*, 121–125.
138. Squartini, F., & Bistocchi, M. (1977). Bioactivity of C3H and RIII mammary tumor viruses in virgin female BALB/c mice. *Journal of National Cancer Institute, 58*, 1845–1847.
139. Squartini, F., Rossi, G., & Paoletti, I. (1963). Characters of mammary tumours in BALB/c female mice foster-nursed by C3H and RIII mothers. *Nature, 197*, 505–506.
140. Stylianou, S., Clarke, R. B., & Brennan, K. (2006). Aberrant activation of notch signaling in human breast cancer. *Cancer Research, 66*, 1517–1525.

141. Takahashi, K., Mitsui, K., & Yamanaka, S. (2003). Role of ERas in promoting tumour-like properties in mouse embryonic stem cells. *Nature*, *423*, 541–545.
142. Telesnitsky, A., & Goff, S. P. (1997). Reversed transcription and the generation of retroviral DNA. In J. M. Coffin, S. H. Hughes, & H. E. Varmus (Eds.), *Retroviruses* (pp. 121–160). Cold Spring Harbor Laboratory Press (NY), USA.
143. Theodorou, V., Boer, M., Weigelt, B., Jonkers, J., Van der Valk, M., & Hilkens, J. (2004). Fgf10 is an oncogene activated by MMTV insertional mutagenesis in mouse mammary tumors and overexpressed in a subset of human breast carcinomas. *Oncogene*, *23*, 6047–6055.
144. Theodorou, V., Kimm, M., Boer, M., Wessels, L., Theelen, W., Jonkers, J., et al. (2007). MMTV insertional mutagenesis identifies genes, gene families and pathways involved in mammary cancer. *Nature of Genetics*, *39*, 759–769.
145. Touw, I. P., & Erkeland, S. J. (2007). Retroviral insertion mutagenesis in mice as a comparative oncogenomics tool to identify disease genes in human leukemia. *Molecular Therapy*, *15*, 13–19.
146. Tsukamoto, A. S., Grosschedl, R., Guzman, R. C., Parslow, T., & Varmus, H. E. (1988). Expression of the int-1 gene in transgenic mice is associated with mammary gland hyperplasia and adenocarcinomas in male and female mice. *Cell*, *55*, 619–625.
147. Uren, A. G., Kool, J., Berns, A., & van Lohuizen, M. (2005). Retroviral insertional mutagenesis: Past, present and future. *Oncogene*, *24*, 7656–7672.
148. Uren, A. G., Kool, J., Matentzoglu, K., de Ridder, J., Mattison, J., van Uitert, M., et al. (2008). Large-scale mutagenesis in p19(ARF)- and p53-deficient mice identifies cancer genes and their collaborative networks. *Cell*, *133*, 727–741.
149. van Amerongen, R., Mikels, A., & Nusse, R. (2008). Alternative wnt signaling is initiated by distinct receptors. *Science Signaling*, *1*, re9.
150. Van Der Valk, M. A. (1981). Tumor incidence of the inbred mouse strains in the Netherlands cancer institute. In J. Hilgers & M. Sluyser (Eds.), *Mammary tumors in the mouse* (pp. 45–115). Amsterdam: Elsevier.
151. Van Nie, R. (1981). Mammary tumorigenesis in the GR mouse strain. In M. Sluyser & J. Hilgers (Eds.), *Mammary tumors in the mouse* (pp. 202–266). Amsterdam: Elsevier/North-Holland Publishing.
152. van Nie, R., & Dux, A. (1971). Biological and morphological characteristics of mammary tumors in GR mice. *Journal of National Cancer Institute*, *46*, 885–897.
153. van Nie, R., & Verstraeten, A. A. (1975). Studies of genetic transmission of mammary tumour virus by C3Hf mice. *International Journal of Cancer*, *16*, 922–931.
154. van Nie, R., Verstraeten, A. A., & de Moes, J. (1977). Genetic transmission of mammary tumour virus by GR mice. *International Journal of Cancer*, *19*, 383–390.
155. Varmus, H. E., Quintrell, N., Medeiros, E., Bishop, J. M., Nowinski, R. C., & Sarkar, N. H. (1973). Transcription of mouse mammary tumor virus genes in tissues from high and low tumor incidence mouse strains. *Journal of Molecular Biology*, *79*, 663–679.
156. Veltmaat, J. M., Mailleux, A. A., Thiery, J. P., & Bellusci, S. (2003). Mouse embryonic mammogenesis as a model for the molecular regulation of pattern formation. *Differentiation*, *71*, 1–17.
157. Vogt, V. M. (1997). Retroviral virions and genomes. In J. M. Coffin, S. H. Hughes and H. E. Varmus (Eds.), *Retroviruses* (pp. 27–69). Cold Spring Harbor Laboratory Press (NY), USA.
158. Wang, Y., Holland, J. F., Bleiweiss, I. J., Melana, S., Liu, X., Pelisson, I., et al. (1995). Detection of mammary tumor virus env gene-like sequences in human breast cancer. *Cancer Research*, *55*, 5173–5179.
159. Wang-Johanning, F., Frost, A. R., Johanning, G. L., Khazaeli, M. B., LoBuglio, A. F., Shaw, D. R., et al. (2001). Expression of human endogenous retrovirus k envelope transcripts in human breast cancer. *Clinical Cancer Research*, *7*, 1553–1560.
160. Watson, C. J., & Khaled, W. T. (2008). Mammary development in the embryo and adult: A journey of morphogenesis and commitment. *Development*, *135*, 995–1003.

161. Wei, Q., Yokota, C., Semenov, M., Doble, B., Woodgett, J., & He, X. (2007). R-spondin1 is a high affinity ligand for LRP6 and induces LRP6 phosphorylation and beta-catenin signaling. *Journal of Biological Chemistry, 282*, 15903–15911.
162. Wellinger, R. J., Garcia, M., Vessaz, A., & Diggelmann, H. (1986). Exogenous mouse mammary tumor virus proviral DNA isolated from a kidney adenocarcinoma cell line contains alterations in the U3 region of the long terminal repeat. *Journal of Virology, 60*, 1–11.
163. Weng, A. P., Ferrando, A. A., Lee, W., Morris, J. P., Silverman, L. B., Sanchez-Irizarry, C., et al. (2004). Activating mutations of NOTCH1 in human T cell acute lymphoblastic leukemia. *Science, 306*, 269–271.
164. Wong, G. T., Gavin, B. J., & McMahon, A. P. (1994). Differential transformation of mammary epithelial cells by Wnt genes. *Molecular and Cellular Biology, 14*, 6278–6286.
165. Wood, L. D., Parsons, D. W., Jones, S., Lin, J., Sjoblom, T., Leary, R. J., et al. (2007). The genomic landscapes of human breast and colorectal cancers. *Science, 318*, 1108–1113.
166. Wu, X., Li, Y., Crise, B., & Burgess, S. M. (2003). Transcription start regions in the human genome are favored targets for MLV integration. *Science, 300*, 1749–1751.
167. Yanagawa, S., Kakimi, K., Tanaka, H., Murakami, A., Nakagawa, Y., Kubo, Y., et al. (1993). Mouse mammary tumor virus with rearranged long terminal repeats causes murine lymphomas. *Journal of Virology, 67*, 112–118.
168. Yanagawa, S., Tanaka, H., & Ishimoto, A. (1991). Identification of a novel mammary cell line-specific enhancer element in the long terminal repeat of mouse mammary tumor virus, which interacts with its hormone-responsive element. *Journal of Virology, 65*, 526–531.
169. Yoshimura, N., Sano, H., Hashiramoto, A., Yamada, R., Nakajima, H., Kondo, M., et al. (1998). The expression and localization of fibroblast growth factor-1 (FGF-1) and FGF receptor-1 (FGFR-1) in human breast cancer. *Clinical Immunology and Immunopathology, 89*, 28–34.
170. Zhu, Q., & Dudley, J. P. (2002). CDP binding to multiple sites in the mouse mammary tumor virus long terminal repeat suppresses basal and glucocorticoid-induced transcription. *Journal of Virology, 76*, 2168–2179.

Chapter 4
Chicken Models of Retroviral Insertional Mutagenesis

Vladimír Pečenka, Petr Pajer, Vít Karafiát, and Michal Dvořák

The concept of transposon tagging or insertional mutagenesis as a strategy for fishing out genes connected with the phenotype of interest was emerging since the early 1980s. Study of genetic basis of tumorigenesis is one of the fields where insertional mutagenesis proved to be exceptionally powerful. Crucial elements of this experimental approach have been retroviruses whose unique properties have revolutionized the work in the field of oncogenesis. Retroviruses contributed to our knowledge of tumor formation in two ways. First, some of them transduce oncogenes—mutants of normal cellular genes with an oncogenic potential. And it was the comparison of viral and cellular alleles of these genes that allowed comprehending the principles of oncogenic activation of genes. Second, retroviruses not carrying oncogenes can induce tumors by affecting host genes. Through integration of their proviral DNA into chromosomes they can activate tumorigenic potential of oncogenes or inactivate tumor suppressor genes. The mechanism is referred to as oncogenesis by insertional mutagenesis. The insertional mutagenesis by retroviruses is very efficient. Perhaps each locus of a host genome can be hit by the provirus insertion in many cells of an infected tissue. If any of these insertions or their combinations incites malignant transformation, the touched cell outgrows and can give rise to a tumor. Affected host gene loci can be easily identified since they are tagged by integrated proviral sequences.

Chicken retroviruses and chickens as experimental animals laid the grounds to the entire field. Chicken B–lymphoma induced by the avian leukosis virus (ALV) was the first tumor type where this mechanism of tumorigenesis was reported in 1981. In these tumors the cellular oncogene *c–myc* was found activated by a provirus integration. In coming years, these pioneering works were followed by further papers that included other retroviruses and animal models and found other insertionally activated oncogenes in other tumor types, thus establishing insertional mutagenesis as a new paradigm in retroviral oncogenesis. The speed of further data accumulation was increasing with continuous improvement of the techniques

V. Pečenka (✉)
Department of Molecular Virology, Institute of Molecular Genetics, Prague, CZ–14220, Czech Republic
e-mail: vladimir.pecenka@img.cas.cz

A.J. Dupuy, D.A. Largaespada (eds.), *Insertional Mutagenesis Strategies in Cancer Genetics*, DOI 10.1007/978-1-4419-7656-7_4,

Table 4.1 Insertionally activated oncogenes in chicken cancer models

Tumor	Virus	Oncogene	Gene type	Frequency of activation (%)	References
B–lymphoma	RAV	***c–myc***	transcription factor	> 95	[126]
B–lymphoma	RAV	***Bic/miR–155***	miRNA	14 (50% in metastasis)	[23]
B–lymphoma (short latency)	RAV, EU–8	***c–myb***	transcription factor	~100	[23] [119]
B–lymphoma (short latency)	RAV	***tert***	telomerase catalytic subunit	~14	[165]
erythoblastosis	RAV	***c–erbB/EGFR***	receptor tyr–kinase	> 95	[92] [123]
nephroblastoma	MAV	***foxP1***	transcription factor	~6	[108]
nephroblastoma	MAV	***plag1***	transcription factor	~5	[108]
nephroblastoma	MAV	***twist***	transcription factor	~3	[108]
nephroblastoma	MAV	***c–Ha–ras***	G–protein	~2	[108]
lung sarcoma	MAV	***frk***	tyrosine kinase	> 95	[106]
hepatocarcinoma	MAV	***c–Ha–ras***	G–protein	~39	Pajer et al., in preparation
hepatocarcinoma	MAV	***c–erbB/EGFR***	receptor tyr–kinase	~32	Pajer et al., in preparation
hepatocarcinoma	MAV	***c–ron/MST1R***	receptor tyr–kinase	~11	Pajer et al., in preparation
hepatocarcinoma	MAV	***c–met/HGFR***	receptor tyr–kinase	~7	Pajer et al., in preparation

for isolating provirus-flanking sequences (inverse PCR and related techniques) and for their analysis (automatic sequencing machines). The final explosion of data started after the complete genome sequences of individual model organisms had become available. From that point it has been a relatively easy task to assort individual provirus insertion sites along the chromosomes, to pick up common sites of integration and associate them with suspect genes.

In this chapter we will describe several chicken models of tumorigenesis through retroviral insertional mutagenesis that have been thoroughly investigated (for the overview, see Table 4.1). More space will be given to the recently advanced models: chicken nephroblastomas, lung sarcomas, liver carcinomas and the industasis phenomenon. We will focus on unique features of the chicken models including mechanistic details of gene activation by defective proviruses. Insertional inactivation of tumor suppressor genes will not be discussed since it has not been proven in chicken system so far. For general issues we refer to other parts of this book and to the reviews [75, 150]. In the end, we will discuss the future of chicken models in the era of high throughput oncogene screening in mouse models and large scale sequencing of human cancer genomes.

4.1 The Beginning of the Story: the Case of Chicken Bursal Lymphomas

Bursal lymphoma (malignant lymphoma of the bursa of Fabricius named also lymphoid leukosis or lymphomatosis) is the most common neoplasia of domestic chickens. The viral etiology of bursal lymphomas has been known since 1908 when cell-free extract was shown to transmit the neoplasia [41]. In farms, the causative agent, now called avian leukosis virus (ALV), is spread mainly by congenital infection from hens to embryos [16]. Experimentally, the virus can be transmitted by intraperitoneal injection of plasma of infected animal into newly hatched susceptible chickens. Several weeks after infection many microscopic transformed follicles arise in bursa of Fabricius (see also Section 4.7). Within months, one or more macroscopically observable nodules—bursal lymphomas—develop. The animals die after progressive metastatic tumors develop in many organs [4, 28].

ALVs are typical representatives of simple slowly oncogenic retroviruses carrying only viral replicative genes gag, pol and env to which no substantial transforming potential could be ascribed [25]. The first suggestion about the mechanism by which these transforming gene-lacking viruses could induce neoplasia emerged from the study of recombinants between oncogenic ALVs (RAV–1 and RAV–2, Rous associated viruses subgroups A and B) and their nononcogenic homolog (RAV–0, Rous associated virus subgroup E). The oncogenicity of the recombinant always correlated with the presence of U3 region derived from RAV–1 or RAV–2 [31, 149]. The U3 region just precedes the start of viral mRNA transcription at the 5′ end of the integrated DNA provirus. Not surprisingly, sequence analyses have localized typical promoter sequences for RNA polymerase II in the U3 region [164].

Since the U3 region, as part of the long terminal repeat (LTR), is also present at the 3′ end of the provirus, it could drive transcription not only of retroviral genes, but also of downstream located host genes (the mechanism named "promoter insertion", see Fig. 4.1a). Hybrid RNAs containing both viral and cellular sequences have indeed been found in infected cells [122]. The non-oncogenicity of RAV–0

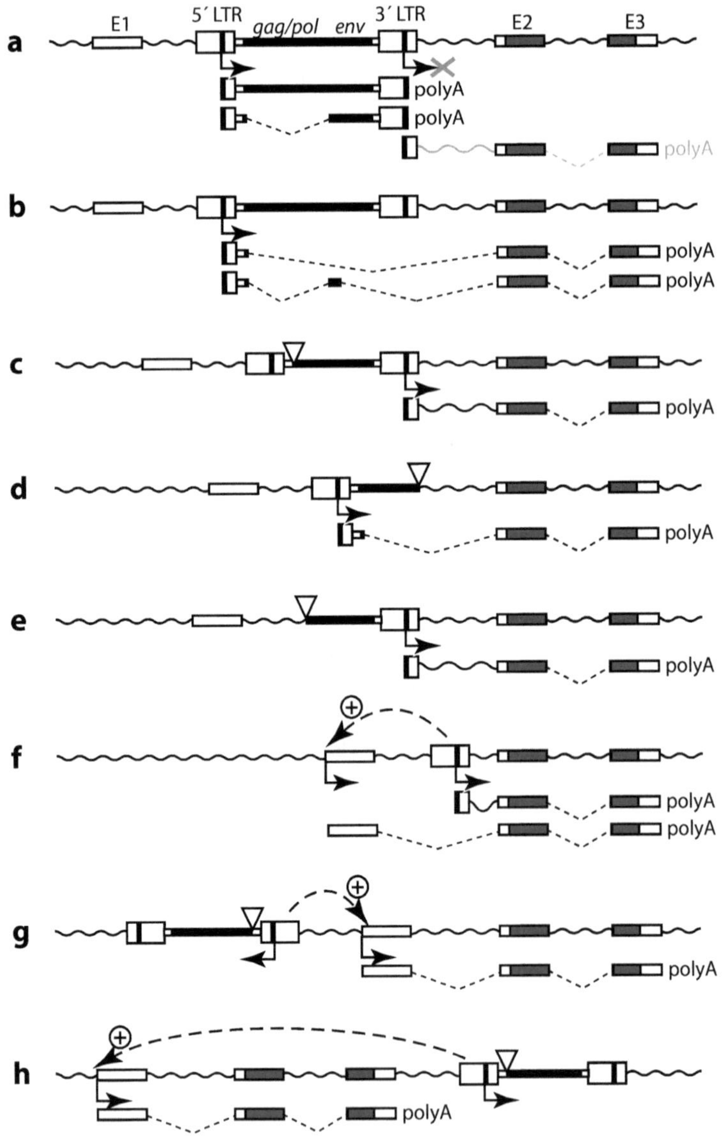

Fig. 4.1 Schemes of gene activation by promoter insertion (a–f) and enhancer insertion (f–h). **a** Nondeleted provirus. Predominant transcripts driven by proviral LTRs are full-length and spliced copies of the provirus initiated at the 5′LTR promoter. Transcription starting in the 5′LTR and proceeding through the 3′LTR arrests the promoter activity of the 3′LTR.

and of recombinants carrying RAV–0-derived U3 could be the consequence of the weakness of RAV–0 promoter which correlated with the slower replication of these viruses [149]. Thus, all the facts were consistent with the hypothesis of insertional activation of an adjacent cellular gene.

The confirmation came when several groups [27, 45, 54, 55, 96, 112] reported frequent provirus integration next to the same discrete cellular sequence in B-lymphomas (denominated as common integration site—CIS). In vast majority of B-lymphomas one of the integrated proviruses was found just upstream of the *c–myc* gene, at the time the already known cellular oncogene. The provirus position and orientation were exactly as required for *c–myc* insertional activation; high levels of hybrid viral/*c–myc* mRNAs were detected in tumors. The same situation was found in chicken B-lymphomas induced by CSV strain of REV [100]. (CSV is another leukemogenic chicken retrovirus lacking an oncogene, unrelated to ALV but classified with mammalian C-type retroviruses).

4.2 Efficient Activation of *C–MYC* Requires Defective Provirus: the Transcriptional Interference and Related Phenomena

More detailed analysis of mechanisms of insertional activation indicated that the original scheme (Fig. 4.1a) was oversimplified. The scheme, where both the 5′LTR

Fig. 4.1 (continued) **b** Nondeleted provirus–readthrough mechanism. Some transcripts initiated at the 5′LTR promoter continue through the weak polyA site in the 3′LTR and are processed at cellular polyA sites. Splicing takes place between the gag splice donor and the cellular splice acceptor. In some instances, a miniexon corresponding to the beginning of *env* is included. **c** Deleted provirus–internal deletion. Deletion of the sequence element downstream of 5′LTR disables 5′LTR promoter activity (see the proposed mechanism in Section 4.2 and Fig. 4.2a). Transcription of downstream sequences is driven by the released 3′LTR promoter. **d** Deleted provirus – 3′LTR deleted. Transcription starts in the 5′LTR. No readthrough is required. The env splice acceptor competing with the cellular splice acceptor is frequently also eliminated. **e** Deleted provirus–5′LTR deleted. Transcription starts in the 3′LTR promoter that was unblocked by 5′LTR deletion. **f** Deleted provirus–single LTR left. Transcription of a downstream cellular gene is driven by proviral promoter (promoter insertion). Alternatively, the authentic cellular gene promoter is stimulated by the proviral enhancer (enhancer insertion). In case the LTR has inserted inside the gene structure both transcription readthrough and splicing take place. **g,h** Enhancer insertion mechanism. Provirus is located downstream of the gene in the same transcriptional orientation or upstream of the gene in the reverse orientation. Transcription starting at the authentic gene promoter is stimulated by an enhancer in the proximate LTR. Also proviruses activating by enhancer insertion carry deletions (e.g. the internal one, as shown here); for details see Section 4.2 and Fig. 4.2b. Only schemes validated experimentally are presented here. Obviously, further schemes combining elements of those shown above can be conceived. Open and black boxes represent noncoding and coding parts of exons, respectively (except for the R region inside the LTRs which, though noncoding, is also black). Wavy lines are introns and non-transcribed regions of genomic DNA, cranked arrows indicate starts of transcription, broken dashed lines indicate splicing; open triangles show the points where deletion has occurred; dashed arrows symbolize stimulation of a promoter by an enhancer and the 3′LTR promoters were supposed to be simultaneously active was in contradiction with the phenomenon of transcriptional interference, originally named promoter occlusion [1]. Particularly in retroviruses it was shown that an efficient initiation in the upstream promoter (i.e. 5′LTR) and progression through the downstream promoter (i.e. 3′LTR) prevents assembling of an initiation complex at the downstream promoter and abolishes its activity [33].

The distinction between 5′ and 3′LTR is further increased by the presence of positive regulatory elements outside the LTRs. In ALV, one such accessory element, called *gag* enhancer, is located in 5′ part of the *gag* gene and acts preferentially on 5′LTR [3, 129]. In CSV, other accessory element lies immediately downstream of 5′LTR but does not behave like an enhancer—it is operative only when located downstream of the promoter and only if present in the proper orientation [10]. That suggests the element performs on the transcript rather than on DNA level, alike TAR element in HIV [25]. Since both elements selectively activate promoter in 5′LTR, they, indirectly, through transcriptional interference, reinforce 3′LTR promoter inhibition.

Accordingly, when the structure of LTR-initiated transcripts in unselected population of ALV-infected cells was carefully examined, less than 2% of them have been found to initiate in 3′LTRs [58]. There was ca 15% of retroviral transcripts that contained both viral and cellular sequences but these hybrid RNAs have been generated by initiation at 5′LTR and by transcription passing through the leaky 3′LTR polyA signal into adjacent host sequences; polyadenylation took place at a distant cellular poly A signal [58]. Hence, the expression of *c–myc* oncogene in B-lymphomas could be, in theory, mediated by hybrid readthrough RNA instead of 3′LTR-initiated RNA. The readthrough RNA would have to be further processed by splicing; if unprocessed, it would code for the viral gag/pol proteins only (see Fig. 4.1b). The splicing would join first six codons of gag with the *c–myc* coding exon; incidentally, such splicing would properly, without a frameshift, join gag and *c–myc* reading frames. However, the readthrough mechanism of *c–myc* activation was not observed in B-lymphomas (see below). Why this mechanism is not effective in the case of *c–myc* in B-lymphomas (while it works quite well with other oncogenes, see next sections) can be explained by the fact that it involves two relatively inefficient steps: 3′LTR readthrough and splicing to *c–myc* second exon splice acceptor while skipping the env splice acceptor. Apparently, the *c–myc* expression levels obtained would not reach the threshold level required for chicken B-lymphomagenesis [109].

Final resolution of activation mechanisms followed from detailed analyses of ALV/*c–myc* arrangement in B-lymphomas. Extensive mapping showed that none out of nearly one hundred of *c–myc* activating proviruses under study was complete [49, 111, 126]. The most frequent defects were internal deletions from 0.5 to 6 kb long that included sequences in the 5′region of the provirus immediately downstream of the LTR. Such deletions result from reverse transcriptase switching the template during the provirus synthesis [168]. Other defects were 5′end or 3′end deletions, some of them incurred during or after the integration since the deletion extended into adjacent cellular sequences. Sometimes nearly all proviral sequences were found deleted leaving only solitary LTR, probably the result of homologous recombination between the two LTRs.

Frequent generation of defects during retrovirus propagation has been well documented [72, 79, 135, 155]. That, however, does not explain why every provirus in *c–myc* locus carries a deletion while majority of other proviruses in the same cells are nondefective. It had to be postulated that only defective proviruses could activate *c–myc* and only cells with such insertion were selected during oncogenesis. Indeed, it can be demonstrated that defects in activating proviruses eliminate obstructions to efficient activation. When 3′ LTR is deleted (Fig. 4.1d), inefficient readthrough step is obviated—all transcripts proceed from 5′LTR into the host sequences; moreover, competing env splice acceptor is often also deleted which makes the splicing from gag donor to the *c–myc* second exon acceptor much more efficient. When 5′ LTR is deleted (Fig. 4.1e) or there is only solitary LTR left (Fig. 4.1f), transcription starts in the remaining LTR whose activity was released from transcriptional interference. In all the above cases the authentic *c–myc* promoter is still present upstream of the inserted LTR but its activity is low and does not interfere with the LTR activity.

Internal deletions in proviruses (Fig. 4.1c) had somewhat unexpected consequences. These deletions completely turned off 5′LTR while 3′LTR became fully active [49]. The very same situation (i.e. disabling 5′LTR by deletion of adjacent downstream sequences) was found in *c–myc* activating proviruses in chicken B-lymphomas induced by CSV [147]. This provokes a question of why 5′LTR becomes completely inactive after losing its downstream accessory elements while 3′LTR that also does not have these elements turns fully active. Presently we can only hypothesize that regulatory interrelations in retroviral provirus are more complex then we have supposed and that any disruption of the provirus integrity can modify activity of both LTRs in a way we are still unable to predict. The existence of additional proviral regulatory elements and interactions including mutual interactions of both LTRs has already been suggested by [101].

The explanation can ensue when the original promoter occlusion/transcriptional interference hypothesis is modified to comply with the present-day knowledge of how distant regulatory regions on chromosome execute their effects. Transcriptionaly active regions including retroviral proviruses form dynamic chromatin loop structures. Enhancer and promoter regions, together with all associated regulatory proteins, are attached to each other at the loop base creating so called active chromatin hub. After a pioneer round of transcription, a terminator region is also juxtaposed to the promoter. In this way a transcriptional unit is demarcated and recycling of transcriptional apparatus in this unit is facilitated [35, 40, 116, 159]. In the integrated provirus, however, the situation is more intricate: there are two identical sets of enhancer, promoter and terminator (i.e. 5′LTR and 3′LTR), preceded by original *c–myc* promoter plus enhancer and followed by original *c–myc* terminator. Thus, several alternative looping structures may form; the final structure is a result of mutual competition for complex formation between individual elements. We suggest that accessory element in 5′part of the complete provirus assist preferential joining of 5′ and 3′LTRs into the loop, presumably by itself forming part of the complex

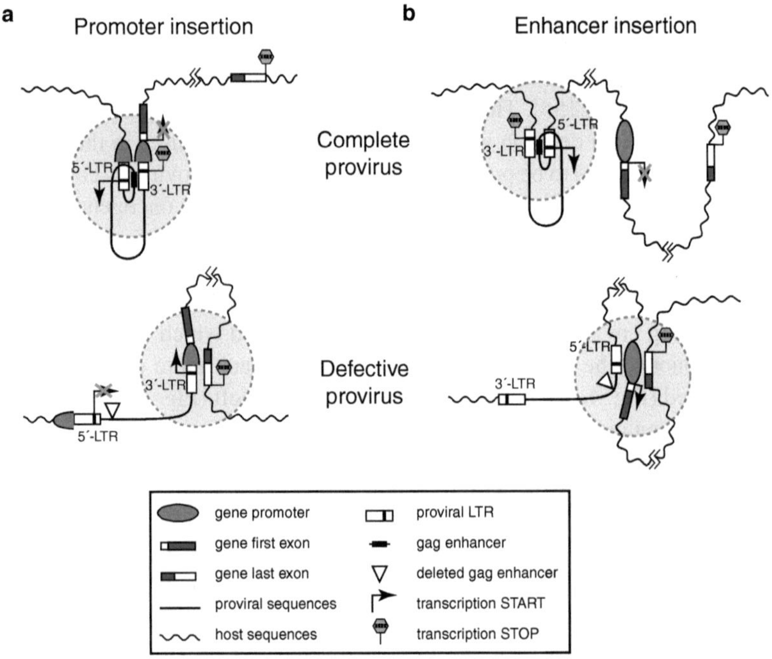

Fig. 4.2 Organization of transcription units controlled by promoters and enhancers of nondeleted and internally deleted integrated proviruses: A hypothesis. **a** Promoter insertion. The complete integrated provirus forms an isolated highly active transcription unit through the formation of a compact chromatin loop structure (active chromatine hub–encircled *gray area*) that is stabilized by interaction of the accessory gag enhancer with both LTRs. In this loop the transcriptional machinery is permanently recycled from the 3′LTR to the 5′LTR. When the gag enhancer is deleted (defective provirus), the proviral loop is destabilized, 5′LTR turns inactive and 3′LTR can form the active transcriptional loop with downstream sequences. **b** Enhancer insertion. The complete provirus forms the active transcriptional unit in which the proviral enhancer is engaged, as in **a**. Therefore the proviral enhancer exerts no influence on regulation of neighboring genes. However, when the provirus chromatine loop is destabilized by deletion (defective provirus), the enhancer is free to interact with neighboring host sequences and stimulates the authentic host gene promoter. All the contacts are mediated by protein complexes associated with promoter, enhancer and terminator regions (not shown in this schematic drawing)

(Fig. 4.2a). The accessory element would thus share some functional properties with the promoter targeting sequence (PTS) described in Drosophila which promotes association of a specific enhancer with a specific distant promoter even when an insulator/enhancer blocker element is present between them [80, 170].

The 5′LTR—3′LTR chromatine hub may be arranged so that an initiation complex is formed only in 5′LTR promoter; however, the existence of the loop itself is sufficient to determine which of the two proviral promoters remains active. When transcription starts in 3′LTR, the transcription complex leaves the loop and new round of promoter selection begins; when transcription starts in 5′LTR, the promoter in 3′LTR is silenced by transcriptional interference and transcription machinery is permanently recycled from 3′LTR back to 5′LTR. In such arrangement, enhancers in both LTRs may stimulate transcription from a single, 5′LTR promoter—the effect observed by [101]. When internal proviral accessory element is deleted as it is the case in most of the *c–myc* activating proviruses the proviral transcriptional unit is destabilized and other looping structures preferentially form including the 3′LTR—*c–myc* polyA loop that demarcates new, highly active c–myc transcriptional unit (Fig. 4.2a).

In several tumors [49, 111, 126], the provirus was located upstream of *c–myc* in the reverse transcriptional orientation or downstream of *c–myc* in the same transcriptional orientation (Fig. 4.1g and 1h). Such arrangement did not conform with the promoter insertion model as transcription starting from any provirus LTR could not proceed to *c–myc* sequences. Nevertheless, substantially increased levels of the *c–myc* mRNA have been found. Its synthesis started at the authentic *c–myc* promoter the activity of which was, apparently, stimulated by juxtaposed powerful enhancer in proviral LTR. This configuration has been denominated "enhancer insertion". It has been reported that also isolated LTR can activate by the enhancer insertion mechanism [81]: in two B-lymphoma lines carrying a single LTR insertion in the first *c–myc* intron in the same transcriptional orientation the LTR-contained enhancer stimulated transcription from the original *c–myc* promoter while the LTR-contained promoter was silenced by transcriptional interference (Fig. 4.1f).

Also proviruses activating *c–myc* by enhancer insertion have been shown to carry internal deletions [49, 126]. That is not surprising when we accept the model presented in previous paragraphs: to be able to attach and stimulate *c–myc* promoter, the relevant proviral enhancer must first be released from the complex with other proviral elements. Deletion of accessory elements in 5′ part of the provirus may be the easiest way of such enhancer liberation (Fig. 4.2b). Obviously, previous reasoning presumes that proviral enhancer cannot make a complex with and stimulate two promoters simultaneously. Such assumption is quite acceptable considering very condensed arrangement of the enhancer into less than 100 bp; it is also in compliance with the provirus tendency to form isolated highly active transcriptional unit.

The support for the above model can be found in a way how MLV and FeLV activate oncogenes. They do it predominantly by enhancer insertion mechanism; however, in contrast to ALV, activating MLV and FeLV proviruses are generally not defective [97, 143]. At first sight it may seem to contradict the proposed model. However, the opposite is true. The above model can, in fact, explain one peculiarity of oncogenesis by MLVs and FeLVs: absolute majority of activating proviruses carry enhancer duplications or even higher level multiplications. For example, lymphomagenesis in some mice strains like AKR relies on spontaneous genesis of oncogenic recombinants (named MCF, mink cell focus-forming virus) from endogenous MLV proviruses present in the strain. Indispensable component of the oncogenic MCF recombinant is a presence in LTR of two copies (in tandem) of strong enhancer acquired from an endogenous provirus in which only single copy of the enhancer is present [143]. Highly lymphomagenic exogenous

MLV strains isolated from naturally occurred tumors, like Mo–MLV, already contain duplicated enhancer [25]. In tumorigenic exogenous FeLVs there is only a single copy of an enhancer; however, in induced T-lymphomas, where *c–myc* is activated by the enhancer insertion mechanism, all activating proviruses carry tandem enhancer duplication acquired always de novo during infection of the animal [97].

The enhancer duplication was shown to have only mild effect on virus replication and provirus transcription. When single-enhancer MLV or FeLV viruses are propagated in culture the anticipated variants with duplication do not arise (or, more accurately, are not positively selected for). Hence, rather than being important for the virus spread in the animal, the duplication is required for the mechanism of oncogenesis itself [97, 143]. We suggest that presence of tandem copies of an enhancer enables provirus 5′LTR to form two chromatin loops simultaneously. While one enhancer is engaged in the formation of proviral transcriptional unit the other one may contact oncogene promoter; no provirus defect is needed for the oncogene activation.

To summarize, distinct ways of how to make provirus capable of strongly affecting host genes are being employed during lymphomagenesis by avian versus mammalian retroviruses. In the ALV model, compactness of the proviral transcriptional unit is broken by different types of provirus deletions, mostly by deletion of the accessory element outside the LTR. Transcription of proximate host genes is then driven by the promoter of defective provirus (mechanism of promoter insertion). In MLV and closely related FeLV, however, no internal accessory elements have been reported. It is possible that compactness of the MLV and FeLV transcriptional units is not dependent on internal accessory elements; thus it cannot be so easy disrupted by provirus deletion. Instead, another way to generate strongly activating proviruses is realized: enhancer multiplication that capacitates provirus to boost transcription of neighboring genes from the genes' own promoters at a distance (mechanism of enhancer insertion). Accordingly, while the mechanism of oncogene activation by enhancer insertion is found only exceptionally in chicken models, it is the dominating mechanism in MLV and FeLV models.

In this context it would be interesting to find out how gene activation works in MMTV model, where enhancer insertion is also the prevalent mechanism. Though the data concerning structural changes in oncogene-activating MMTV proviruses are scarce, it seems that in mammary carcinomas neither deletion nor enhancer duplication takes place [102] while in variant MMTV-induced T-lymphomas enhancer multiplications may play an important role [6, 14]

For the sake of completeness we have to mention additional possible effects of provirus insertion that have not been discussed above. First, an important part of insertional activation mechanism may be separating negative regulatory elements like silencers from the oncogene promoter. Eventually, cellular transcription pausing sites may be shifted away from the transcription unit by the provirus insertion. Presence of negative elements is easy to imagine in cases where the activated gene has been completely inactive before provirus integration. These mechanistic details, however, are commonly not addressed by the investigators. Second, the range of effects incited by the provirus integrating into the gene structure is not limited to the level of transcription. For example, any change in mRNA structure (truncation, fusion with viral sequences, activation of cryptic promoters and splicing or polyadenylation sites etc.) may have profound impact on its export from the nucleus, stability, and rate of translation. In mouse models, the most common mechanism of this type may be the frequent truncation of 3′untranslated regions containing destabilizing AT rich sequences as well as target sequences for miRNAs which downregulate both stability and translatability of affected mRNA. Such effects during insertional activation have occasionally been suspected or even documented, see e.g. [20, 61, 132–134, 152], but only recently have also been more thoroughly analyzed [34].

4.3 Readthrough Activation of *C–MYB* in B-lymphomas: the Case of "Superactivating ALV"

When 10–14 day old embryos, instead of newly hatched chickens, were intravenously injected with ALV, some of them developed, primarily in livers, highly aggressive B-lymphomas after latency of only several weeks (from now on these tumors will be called short latency lymphomas as opposed to long latency lymphomas arising after infection of newborn chicks—see Section 4.2). In most of short latency lymphomas ALV provirus was detected around (upstream, inside or downstream) the first exon of the *c–myb* gene, a cellular progenitor of retrovirally transduced oncogene *v–myb* [23, 117, 118]. The proviruses had no defects and were integrated in the same transcriptional orientation as the *c–myb* gene. Chimeric ALV/*c–myb* mRNAs were synthesized by a typical readthrough mechanism (see Section 4.2 and Fig. 4.1b): they were initiated in 5′LTR, polyadenylated at *c–myb* polyA site and processed by splicing the ALV gag donor to the *c–myb* second exon acceptor. Gag and *c–myb* were not coded by the same reading frame in this message. Moreover, gag start codon was followed by three in frame stop codons at the beginning of *c–myb*; translation of *c–myb* relied on reinitiating at the codon 21. The produced slightly truncated protein was strongly oncogenic, in contrast to the wt protein [63].

Profound effect of infection timing suggests that target cells for short latency lymphomas are different from target cells for long latency lymphomas (presumably prebursal stem cells versus bursal stem cells) and that the first ones are absent in chickens after hatching [118]. In consequence, different target cells are sensitive to activation of different oncogenes (*c–myb* versus *c–myc*). Consistently, no lymphomas develop when infection of the bursa is postponed to the time when bursal stem cells are no more present (ca 3 weeks after hatching), even in animals with subsequent strong lifelong viremia [15].

The frequency of short latency lymphomas and *c–myb* activations were considerably increased when specific recombinant ALV strains were used for infection.

In these "superactivating" strains an element in the 5′part of gag gene, called negative regulator of splicing (NRS), was knocked out by either deletion or mutation [8, 71, 119, 136, 138]. NRS has two roles in ALV both of which help to fulfill peculiar requirements of retroviral replication: to leave substantial proportion of retroviral RNA unspliced and, at the same time, to ensure polyadenylation at the 3′end of such RNA (which is generally coupled to splicing); for details see [24, 84]. To this end NRS first acts as a pseudo-splice donor, forming nonproductive splicing complexes with downstream acceptors thus blocking them from splicing to authentic donor sites. Second, by binding factors that interact with polyadenylation machinery, NRS helps to recruit and stabilize processing complexes at the 3′LTR polyA site. The second effect is dependent on the first one, since only the formation of an abortive splicing complex between NRS and downstream splicing acceptor brings NRS close enough to 3′LTR polyA site to exert its effect on polyadenylation [84].

As shown by [93] on RSV model, deletion of NRS has serious consequences to viral RNA processing. First, 3′LTR readthrough is strongly elevated (up to over 50%). Second, while the frequency of splicing from the gag donor to the env acceptor is, surprisingly, little affected, the splicing to a more distant acceptor is boosted ca four times. Similar (though less conspicuous) effects were observed for "superactivating" ALVs with the mutated NRS [105, 138]. Thus, when such ALV integrates upstream of a cellular gene, the spliced readthrough mRNA coding for the cellular gene is produced with much higher efficiency compared to wt ALV.

4.4 Insertional Activation Can Be Accompanied by Extensive Alterations of the Oncogene Structure and by Formation of an Oncogene-Transducing Virus: the Case of Chicken Erythroblastosis

Besides short- or long-latency B-lymphomas, chickens infected with ALV can infrequently develop other types of neoplasia (see Section 4.8). Among them, the most thoroughly analyzed was erythroblastosis, a disease characterized by uncurbed proliferation of erythroblasts with arrested differentiation. In sensitive chicken strains high amounts of erythroblasts appear in blood ca 2 months after infection with ALV; at the same time they start to massively infiltrate spleen and liver causing rapid death of the animal. Paradoxically, erythroblastosis is often accompanied by anemia since feedback mechanisms triggered by the flood of transformed erythroblasts block further development of nontransformed red blood cell precursors.

In 1983, recurrent insertional activation of the *c–erbB* protooncogene (or *EGFR*—epidermal growth factor receptor gene) in chicken erythroblastosis was first reported [46]. Further papers [50, 92, 99, 123] described peculiar molecular details of this activation. In all analyzed cases, the full-length ALV provirus was integrated in the same intron in the middle of *c–erbB* gene in the same transcriptional orientation. High levels of *c–erbB* specific mRNAs were produced. The mechanism of activation was much the same as activation of *c–myb* in short latency B-lymphomas (Section 4.3): readthrough transcript starting in 5′LTR and polyadenylated at one of the two authentic *c–erbB* polyA sites was processed by splicing that brought *c–erbB* coding sequences close to the start of mRNA (Fig. 4.1b). There were, however, two differences. First, mRNA processing was more complex. An alternative short exon demarcated by env splice acceptor and cryptic splice donor 159 bp downstream was included in high proportion of messages. Second, the extent of oncogene truncation was much more extensive then in the case of *c–myb*. The protein encoded by the spliced mRNAs consisted of 6 AA of gag (plus 53 AA of env in some messages) fused to the C-terminal part of c–erbB. In both protein versions the N-terminal part—the EGF binding extracellular domain—was missing. Such truncation is known to result in a constitutive kinase activity of the receptor [77]. It was suggested that the 53 AA+ version of activated c–erbB may be the major mediator of oncogenesis. The alternative short exon codes for a signal peptide that normally targets env protein for membrane translocation. This function was retained also in a fusion with truncated c–erbB, where it substituted for missing authentic c–erbB signal peptide. During the processing the 53 AA^+ protein had the signal sequence cleaved off, was fully glycosylated and exposed on the cell surface, in contrast to the 53 AA^- version, which was only partially glycosylated and was not exposed on the cell surface [85].

The frequency of erythroblastosis is just several percent in most chicken strains including outbred flocks; there are however inbred strains that develop ALV-induced erythroblastosis with penetrance ca 80%. This sensitivity is a dominant trait not connected with the ability of ALV to activate *c–erbB*, but rather determining sensitivity of the animal to such activation [127].

In high proportion of tumors the situation was obscured by formation of *c–erbB*-transducing recombinant virus. While oncogene transduction is generally a rare event [25], highly leukemogenic viruses carrying truncated version of *c–erbB* can be isolated from up to 50% of chickens with ALV-induced erythroblastosis [92, 124]. That has made erythroblastosis an attractive model and rich source of material for studying mechanisms of transduction. For nearly all other isolates of acutely transforming retroviruses the original tumor where the virus has originated is not available and the virus has further evolved during repeated passages since then. Here, by comparing sequences of the transducing viruses with sequences of the insertionally activated *c–erbB* genes and fusion mRNAs synthesized in the tumors it was possible to confirm basic scheme of transduction as suggested earlier. Details of the mechanism are beyond the scope of this book; for them, see [25, 59, 92, 124, 128, 146]. The important point is that the first step of transduction is the synthesis of hybrid mRNA containing both the retroviral and oncogene sequences, i.e. the transduction is preceded by an insertional mutagenesis. Hence it is not surprising that the sites of truncation in insertionally activated *c–erbB* and in the transduced *v–erbB* precisely coincided.

4.5 Chicken Nephroblastomas Induced by MAV: Complex Model Suitable for High Throughput Oncogene Screening

In previous chapters we have described mechanisms of insertional mutagenesis by retroviruses in hematopoietic cells, which led to the identification of several oncogenes. Each of the models has been dominated by a single CIS. Hematopoietic malignancies are generally characterized by a less diverse pattern of mutated genes than solid tumors. Solid tumors, especially carcinomas, display high complexity and variability of karyotypic alterations, which seriously complicates the search for oncogenes and tumor suppressor genes in these tumors [94]. The complexity probably results from the complex homeostatic control of cells in solid tissues which must be overcome. Thus, finding appropriate model tumors which would enable a large-scale search for solid tumor oncogenes and tumor suppressor genes has been a challenge. The models based on MAV retroviruses are among the successful ones.

MAV–1 and MAV–2 (myeloblastosis-associated viruses 1 and 2) are simple slowly oncogenic retroviruses classified with ALVs. They differ by their env gene (and, consequently, by host specificity), MAV–1 belonging to subgroup A, MAV–2 to subgroup B. They are highly homologous to other ALVs except one region: the U3 part of LTR, the region containing proviral enhancer/promoter and playing crucial role in insertional activation (see Sections 4.1 and 4.2). It is this MAV-specific region that confers very distinct oncogenic potential on MAV retroviruses [67]. When MAV–1 or MAV–2 are injected into 12 day old embryos or newly hatched chicks they induce nephroblastomas with efficiency 70–100% depending on chicken strain. Multiple tumors become visible on kidney of an infected animal as early as 2 months after infection and they rapidly grow into massive size [68, 140, 158, 160].

The chicken nephroblastomas are embryonic tumors derived from nephrogenic blastema cells which persist in the newborn kidney for several days after hatching. They consist mainly of undifferentiated mesenchyme and aberrant differentiating epithelial renal elements reminiscent of developing kidney. They are comparable to human nephroblastomas—Wilms' tumors both by their histology and presumed target cells of tumorigenesis [7, 11, 56, 62].

First attempts to identify CISs in chicken nephroblastomas had only limited success. Two groups reported two different loci hit by proviral integration in chicken nephroblastomas—*Ha–ras* [160] and *nov/ccn3* [68]. Although deregulation of these genes in the chicken or mammalian cells have been proven to result in a transformed phenotype, each of them have been found hit only in a single case and therefore they could not be classified as CIS.

Restriction mapping of integration sites in clonal nephroblastomas using Southern blot analyses revealed unexpectedly complex pattern from which existence of CISs could not be definitely inferred [114]. Three interpretations have been put forward. Firstly, there may be a major CIS but it represents region too large to be disclosed by restriction mapping on the background of several further integrated proviruses in each nephroblastoma clone. Secondly, there might be no major CIS but rather numerous less frequent ones. Thirdly, activating proviruses could carry deletions similarly to the situation in B-lymphomas which would invalidate the used mapping strategy.

Later on, an improvement of the techniques for isolation and analysis of provirus-flanking sequences enabled more comprehensive survey of VISs ("Viral Integration Sites") in nephroblastomas. As described in Section 4.2, defective proviruses are very potent mutagens and deletion in the provirus may be a prerequisite for oncogene activation. Frequent occurrence of defective proviruses in nephroblastomas has been documented before [115, 140, 160]. Therefore, to start with, molecular cloning of VISs harboring defective proviruses in six randomly chosen nephroblastomas have been carried out, leading to identification of the insertionally activated gene *twist*. Subsequent screening of a panel of chicken nephroblastomas have shown that the *twist* gene is the first true CIS in these tumors—it was hit in approximately 4% of analyzed tumors [107].

These results strongly favored the hypothesis of numerous rather infrequent CISs in the chicken nephroblastomas. To prove it, high throughput screening of VIS in hundreds of nephroblastoma samples had to be performed. To achieve a maximum coverage of VISs, two independent and overlapping methods for the VIS identification were employed [108]: inverse PCR (iPCR) and LTR-based rapid amplification of 3′cDNA ends (LTR RACE). The iPCR technique provided genomic sequences flanking the integrated provirus. The technique was optimized to the level when flanking sequences of all clonally integrated proviruses could be gel-isolated after single PCR reaction; routine VISs recovery was 90–100% in each tumor. LTR–RACE technique provided sequences of mRNA species containing provirus LTR at its 5′end, including hybrid mRNAs generated by readthrough transcription or by transcription starting in the promoter of a defective provirus (see Section 4.2 and Fig. 4.1). LTR–RACE analysis covered up to 30% of VISs. The lower yield achieved by LTR–RACE is compensated by the fact that found VISs were the most relevant ones—those where high expression of cellular sequences was driven by an integrated provirus. At the same time, the structure of LTR–RACE products revealed details of the mechanism of host gene activation. Importantly, both techniques detected only insertions present within a substantial proportion of tumor cells as opposed to currently used sequencing strategies that record also oncogenically unselected integrations present even in a single cell.

To date more than 1100 unique VISs have been identified in 187 MAV-induced nephroblastoma clones [108] and (Pajer, unpublished). Out of them 69 candidate tumor-related loci (denominated NALs, "nephroblastoma associated loci") have been selected according to any of the following three criteria:

(a) two and more VISs in a close genomic distance; the distance limit was set arbitrary to 20 kb;
(b) two and more VISs within one annotated gene;
(c) VISs from the tumors harboring only a single integrated provirus as demonstrated both by Southern blot hybridization and iPCR; to date only three such samples have been found.

It should be emphasized that not all candidate NALs necessarily have to represent cancer-related genes, especially those hit by a provirus in two samples only. It has been shown that using the sole criterion of multiple insertions into the same locus in retroviral oncogene screens may lead to high proportion of false positives in consequence of retroviral integration preferences [163]. The most relevant loci (those hit in at least four tumors) were the *foxP1*, the *plag1*, the *twist*, and the *c–Ha–ras*. *FoxP1* (forkhead box P1) encodes a widely expressed transcription factor important for heart, lung and lymphocyte development; it was shown to function either as an oncogene or a tumor suppressor gene depending on cell type [76]. *Plag1* codes for a zinc finger transcription factor in humans expressed only in embryonal tissues; its inapropriate activation is responsible for induction of several types of human tumors [151]. *Twist* encodes a basic helix-loop-helix transcription factor regulating epithelial–mesenchymal transition during embryogenesis and metastasis and supporting survival of cancer cells [2, 22].

Analysis of produced mRNAs and provirus alignment in the selected CISs showed that the molecular mechanism of activation was different in each case (see also Fig. 4.3). All *twist*-activating proviruses were integrated in the same transcriptional orientation in a narrow region just upstream of the *twist* initiation codon and most of them carried various deletions or rearrangements, alike the ALV proviruses activating the *c–myc* gene in chicken B-lymphomas (Section 4.2, Fig. 4.1c–f). Very high expression of *twist* mRNA was driven mostly by 3′LTR promoter. Neither normal adult kidney nor nephroblastomas without MAV insertion in the *twist* locus did contain detectable amount of *twist* mRNA while moderate level of *twist* mRNA was detected in the embryonic kidney.

Similarly, in tumors with activated *plag1* gene very high levels of the *plag1* mRNA were found. This RNA was generated by a readthrough mechanism followed by splicing between the MAV gag donor site and the *plag1* second exon acceptor site, much like during *c–myb* activation in short latency lymphomas (Section 4.3, Fig. 4.1b). The hybrid mRNA coded for full-length *plag1* protein.

A strikingly different situation was found in the case of *foxP1*. Proviruses were inserted downstream of the first or second *foxP1* coding exons mostly in the same orientation as the gene's transcription. No gross change in overall expression of *foxP1* mRNA has been observed in any tumor harboring affected allele of the gene. Hybrid MAV/*foxP1* mRNAs, however, could be detected and analyzed using RT–PCR; they coded for the N-terminally truncated FoxP1 proteins. Further analysis of the effect of *foxP1* truncation was complicated by the existence of alternative starting points, alternative exons and alternative polyA signals in *foxP1* gene. Moreover, it has been observed that both the amount and subcellular localization of *FoxP1* protein was highly variable in individual nephroblastoma samples, including those not harboring viral integration in *foxP1* gene.

As to the *c–Ha–ras* activation, provirus insertions have been found in 5′untranslated region of the gene in the same transcriptional orientation. Both hybrid MAV/*c–Ha–ras* mRNAs and normal *c–Ha–ras* mRNAs were observed; the hybrid mRNAs coded for normal c–Ha–ras protein but they were missing 5′untranslated region present in normal *c–Ha–ras* mRNA. The overall level of *c–Ha–ras* mRNA

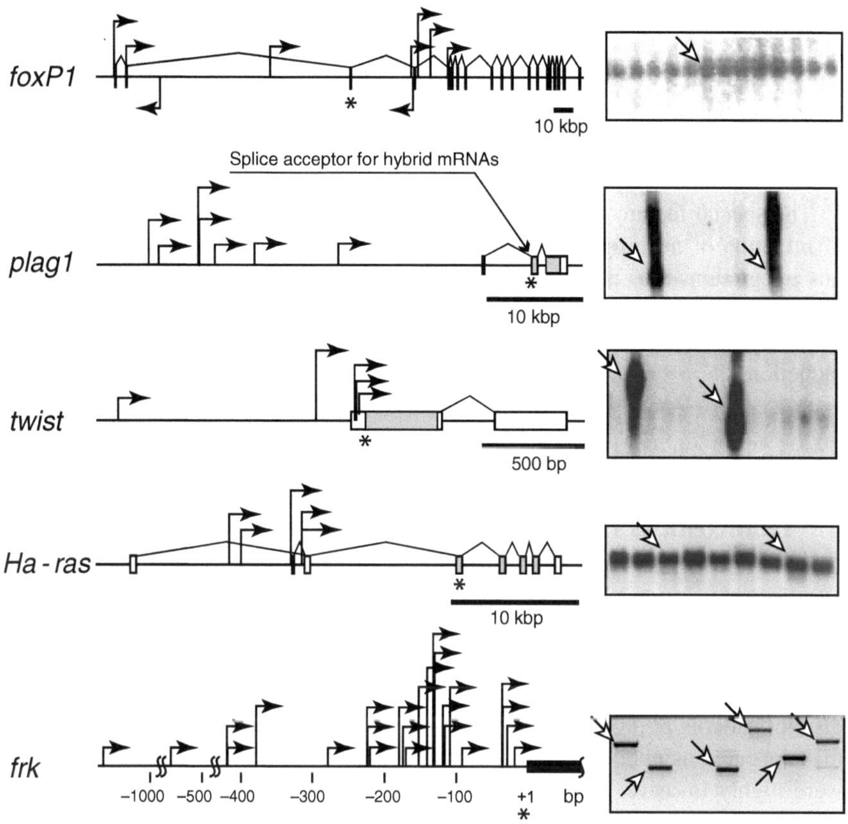

Fig. 4.3 Genomic organization, distribution of insertion sites and mRNA expression in selected CISs. The CISs were identified in chicken nephroblastomas (*foxP1, plag1, twist and Ha ras*) or lung sarcomas (*frk*). Sites and orientations of individual proviral insertions are indicated by *arrows*. Positions of translation initiation are marked by asterisks. Panels to the right show Northern blot hybridization profile of a few representative tumor samples, except for the *frk* locus where RT–PCR profile is shown. The samples harboring a provirus in the given locus are marked by *arrows*

was the same in nephroblastomas with or without insertion in *c–Ha–ras* locus and in the normal kidney. No point mutations, the archetypal mode of *Ha–ras* activation in human and mouse tumors, have been found (Pajer, unpublished). How such "activation", which affects neither mRNA level nor the structure of the produced protein can participate in nephroblastoma induction is not clear. We suggest that deleting 5′untranslated region might result in the increased translation of the hybrid mRNA. Alternatively, *Ha–ras* activation could be important only during the induction stage in nephroblastema cells, the target cells of oncogenesis—for instance, if basal level of *Ha–ras* expression is low in these cells (hit and run hypothesis).

The results draw a highly complex picture of virus-induced mutations in nephroblastomas. Out of 69 candidate NALs only four were hit in a reasonable percentage of samples: the *foxP1* (in about 6% of tumors), the *plag1* (ca 5% of tumors), the *twist* (ca 3% of tumors) and the *c–Ha–ras* (ca 2% of tumors). All the identified NALs together would explain for induction of ca 130 out of 197 nephroblastomas (ca 65%). Evidently, many alternative cancer-related genes (or combinations of them) may be deregulated by the provirus integration in nephroblastomas.

The special feature of the chicken nephroblastoma model is the high efficiency of infection of nephrogenic blastema cells [114, 160]. Each cell acquires numerous integrations thus generating enormous amount of combinations of insertional mutations. These facts establish the chicken nephroblastoma as a perfect model for the study of oncogene/tumor suppressor gene cooperation during multistep tumor induction.

4.6 Extension of Spectrum of MAV-Induced Tumors By Local Homeostasis Perturbation: Lung Sarcomas, Liver Carcinomas and the Industasis Phenomenon

In about 7% of MAV–2-infected animals late lung sarcomas or liver carcinomas developed in addition to nephroblastomas. These tumors were independent primary clonal outgrowths, not metastases originated from the nephroblastoma as shown by distinct patterns of proviral integrations. While nephroblastomas and liver carcinomas appeared as nodules clearly separated from normal tissue, the lung sarcomas were highly invasive: one tumor clone was frequently disseminated into several foci in both lungs and in some animals these foci constituted majority of lung tissue [106]. Presence of considerable amount of tumor cells could even be detected by PCR in macroscopically normal lung tissue samples (Pajer, unpublished). Both types of late tumors would represent new interesting models for oncogene screening if it were not for the crucial drawback: they were too rare to provide enough samples for thorough analysis.

The obstacle of low penetrance was cleared by discovery that injection of MAV–2 producing chicken cells instead of virions intravenously into the embryos or newly hatched chickens changes formerly rare tumors into frequent ones. In the case of lung tumors the process of tumorigenesis was also markedly accelerated—the lung tumors appeared in most animals within 1–2 months after cell injection, prior to nephroblastomas and liver tumors, the latency of which did not change. The cell-assisted tumors did not differ from the rare ones neither by gross morphology nor by histology. Preventing the division of injected MAV–2 producing cells by treatment with mitomycin did not abolish their ability to promote tumor formation. Consistently, the cell-assisted lung and liver tumors were shown to be of the host origin and not to contain detectable amounts of the injected cells [106] and (Pajer et al., in preparation).

Surprisingly, the same tumor-promoting effect was observed when animals were injected first with MAV–2 virus at embryonal day 12 and then with uninfected chicken embryonal fibroblasts 9 days later, shortly after hatching. At that time all tissues of the animal including lungs and liver were already fully infected, suggesting that injected cells do not function only as reservoirs spreading virus in the animal. This conclusion was further supported by the fact that injected cells affected neither the spectrum of infected tissues nor the level of their infection. However, when injections were carried out in the opposite order, i.e. the virus was injected 9 days after the cells, no tumor promotion was observed [106]. This fact suggests that the tumor-promoting effect of the injected cells persists not longer than several days.

To follow the fate of the injected cells, [35S]methionine-labelled embryonal fibroblasts were used and monitored in individual embryonic tissues both by total tissue radioactivity and by paraffin-embedded section autoradiography. The experiments showed a prompt passage of the injected cells from blood into the tissues where they settled as individual stray cells. 20 h after injection the cells were detected in all analyzed embryonic tissues including brain; the highest levels were found in the liver, kidney, and lungs while the residual levels in the blood were very low.

Based on these observations the concept of industasis was proposed—a promotion of fully malignant phenotype of an incipient tumor cell by a stray cell through a disruption of local homeostasis; for details see [106]. It was suggested that the phenomenon of industasis might be the underlying cause of many human multiple primary tumors when a stray cell released by an advanced tumor promotes tumor formation from another, mutated and potentially malignant cell that was kept under control by tissue homeostasis.

Besides its potential impact on our understanding of mechanisms of oncogenesis in general, the industasis fenomenon advanced the model of MAV–2 insertional mutagenesis as it enabled a large scale analysis of lung and liver tumors. Both rare virus-only-induced and frequent cell-assisted tumors have been searched for CISs using the same methods as described in Section 4.5, i.e. iPCR and LTR RACE. Quite dissimilar pattern of CISs was revealed in the lung versus liver tumors; importantly, however, there was no difference between the rare and the cell-promoted tumors [106].

In contrast to nephroblastomas, where numerous low frequency CISs have been found, lung sarcomas were dominated by a single one. In more than 95% lung tumors a provirus insertion was observed in the gene *frk/rak* [106] and (Pajer et al., unpublished); no other CIS was identified. The *frk* (fyn-related-kinase gene) encodes a tyrosine kinase expressed in humans predominantly in epithelial tissues [19]. It was shown to act as a tumor suppressor gene that negatively regulates PI3K/Akt pathway through stabilisation of PTEN, another tumor suppressor and negative regulator of PI3K [13, 167]. Surprisingly, *frk*-knockout mice displayed very mild phenotype and no increase in tumor incidence [21]. The model of chicken lung sarcomas has demonstrated for the first time the oncogenic capability of *frk* overexpression, previously being only suspected [60].

The provirus/*frk* arrangement was much alike provirus/*c–myc* arrangement in B-lymphomas: defective proviruses had integrated in the same transcriptional orientation into a very narrow area within the promoter/5′UTR region in front of the gene's coding sequences. Most defects were internal deletions comprising the enhancer element in 5′part of *gag* (Pecenka et al., in preparation). High levels of MAV–2/*frk* hybrid mRNA initiating in the proviral 3′LTR have been detected in all samples carrying provirus insertion in the *frk* locus. The protein coding sequence was not afflicted in any sample. No *frk* expression was detectable neither in tumors with unaffected *frk* locus nor in the non-tumor lung tissue [106].

In liver tumors, the pattern of CISs was more variable and different from the pattern in kidney or lung tumors (Pajer et al., in preparation) and (Pecenka et al., in preparation). The vast majority of liver tumors carried MAV–2 provirus inserted within one of four genes: *c–Ha–ras* (39% of tumors), *c–erbB/EGFR* (32%), *c–ron/c–stk/c–sea/MST1R* (11%) or *c–met/HGFR* (7%); no other CIS was recorded. The last three genes encode receptor tyrosine kinases expressed in many types of epithelial cells and implicated in several carcinoma classes in humans [47, 104, 156]. Interestingly, they all control, among others, signaling pathways converging on the *c–Ha–ras,* the most frequently activated gene in chicken liver tumors. That suggests that activation of the *c–Ha–ras* signaling is the pivotal disturbance in all MAV–2 induced liver tumors and that insertions into four alternative genes might be just four different ways how to affect *c–Ha–ras* and pathways downstream of it. Consistently, simultaneous activation of two or more of these genes in the same tumor was never observed, though, on mere statistical basis, *c–Ha–ras* should be hit in about one third of tumors with *c–erbB* activation and vice versa. It must be pointed out, however, that the screen did not comprise genes connected with progression steps since all samples analyzed were early stages of tumor development, according to terminology of [73] classified as preneoplastic nodules or adenomas (Pajer et al., in preparation).

Similarly to nephroblastomas, *c–Ha–ras* gene was activated by provirus insertion into the gene's 5′untranslated region; the *c–Ha–ras* coding sequences were preserved. Alike in nephroblastomas, no point mutations have been found. Unlike nephroblastomas, however, insertions resulted in substantial overexpression of hybrid MAV–2/*c–Ha–ras* mRNA initiating in the proviral 3′–most LTR. All activating proviruses have been heavily rearranged including large deletions, duplications and inversions (Pecenka et al., in preparation).

The mode of *c–erbB* activation resembled the situation in ALV-induced erythroblastosis (Section 4.4): all proviruses had integrated into the same intron in the middle of the gene. That resulted in high expression of hybrid mRNA encoding the N-terminally truncated c–erbB/EGFR fused to the first 6 AA of *gag* gene. Contrary to erythroblastosis the *c–erbB*-activating proviruses had been heavily rearranged much alike the *c–Ha–ras*-activating ones. Due to the provirus rearrangement, variant mRNA containing *env* mini-exon described in erythroblastosis model could not be generated.

The *c–ron* activation paralleled the *c–erbB* activation in all aspects described above including the truncation of gene's 5′part that codes for the receptor's extracellular ligand-binding domain. Alike in case of c–erbB, N-terminal truncation confers constitutive kinase activity on c-ron protein [82].

The details of *c–met* activation could not be conclusively established since RNA from these tumors was not available. Based on DNA analysis it seems that the integrated defective proviruses could drive overexpression of the full-length protein—they were located upstream of the gene's initiation codon (in the promoter or 5′untranslated region), in the same transcriptional orientation.

4.7 Evidence of Multistage Cancerogenesis in Chicken Models

Tumors are end products of successive evolution through selection of progressively more malignant cell subclones which is reflected by complex pattern of accumulated mutations of oncogenes and tumor suppressor genes ("cancer genes"). Such complex pattern is typical for human tumors and has also been found in mouse tumor models. In majority of chicken models the mutational pattern seems to be quite simple (see previous Sections); that may, however, be just due to the absence of thorough high throughput analyses. Nevertheless, the multistep nature of oncogenesis in chicken models can be evidenced by studying progressive evolution of tumors.

The examination of early stages of B-lymphoma development (Section 4.1) showed that cells with a proviral integration in the *c–myc* locus appear early after ALV infection. The chicken Bursa of Fabricius is compartmentalized into ca 10 000 follicles which are colonized by B-cell progenitors during late embryonic development [103]. One month after ALV infection of young chicks, up to 100 hyperplastic transformed follicles are observable in bursa. Each of them represents a clonal outgrowth of a different cell with activated *c–myc* [9, 48]. The hyperplastic state is not a consequence of accelerated proliferation but of blockage of lymphoblast differentiation and emigration from bursa [12]. Nearly all transformed follicles disappear during bursa involution, only one or a few (if any) of them progress to the stage of B-lymphoma. Distinct and (to a lesser extent) overgrown clones of cells with activated *c–myc* gene have also been observed 1 month after infection in tissues where primary tumors do not develop (spleen, bone marrow). Thus, *c–myc* activation is the early step in oncogenesis, but it is not sufficient for B-lymphoma formation; further steps must follow. Indeed, when chicken B-lymphomas were searched for additional CISs, the gene *bic* coding for regulatory RNA miR–155 was found insertionally activated in a significant proportion of chicken B-lymphomas, most frequently in metastases together with activated *c–myc* [23, 38]. Undoubtedly, an extensive analysis of chicken B-lymphomas would discover further activated genes cooperating with *c–myc* during tumor development much like in mouse MLV-induced T-lymphomas where many such genes have already been found [91, 142].

A very similar situation was observed during short latency lymphoma development (Section 4.3). Here the *c–myb* gene was found activated in hyperplastic follicles in bursa early after infection. Contrary to long latency B-lymphomas, emigration of the transformed lymphoblasts was not fully blocked. Cells harboring activated *c–myb* also infiltrated neighbouring follicles (without compromising the structure of bursa) so that substantial proportion of bursa was formed by clusters of hyperplastic follicles [98, 117]. While bursa mostly stayed in a hyperplastic state without tumors, the lymphoblasts released from it pervaded liver (plus bone marrow and other organs) and formed B-lymphomas there. Even in chicks which eventually developed no lymphoma, the bursa of Fabricius was mostly overgrown by oligoclonal population of cells with activated *c–myb*. Obviously, lymphoblast clones with activated *c–myb* had a strong advantage when populating bursa, but absolute majority of them did not progress to form tumors. Thus, again, *c–myb* activation is only one event in a multistep process of oncogenesis. The additional steps, however, seem mostly not to involve insertional activation since a high proportion of short latency B-lymphomas contain only a single ALV provirus—the one in the *c–myb* locus [71].

Recently, a frequent insertional activation of telomerase reverse transcriptase gene (*TERT*) was discovered in short latency B-lymphomas [165]. Activation of *TERT* and *c–myb,* however, seem to represent alternative, not cooperating events since tumor clones carrying activated *TERT* did not contain activated *c–myb* and vice versa.

Also the development of erythroblastosis (Section 4.4) shows the same signs of multistep tumorigenesis: multiple distinct clones of cells with insertionally activated *c–erbB* originate and overgrow in hematopoetic tissues but only one or few clones in spleen or bone marrow go on to form a tumor [48]. So far, no search for *c–erbB*-cooperating genes in chicken erythroblastosis was performed. Similarly, no systematic attempts to identify cooperating oncogenes in MAV-induced tumors have been carried out yet.

It must be emphasized that some steps in a multistage retroviral oncogenesis may have no relation to provirus insertions. Spontaneous mutations and chromosomal rearrangements, the major mechanisms operating during human tumorigenesis, may be involved, especially when we realize the huge reservoir of premalignant cells created by the first step of tumorigenesis, i.e. by the insertional activation of an oncogene. However, to discover genes affected by stochastic processes is incomparably more difficult than to pick up genes tagged by integrated proviruses and relevant technologies are very labor-intensive and costly. Consequently, no data are available about possible involvement of point mutations or chromosomal rearrangements in the models described above. Limited searches for Ha–ras mutations in chicken nephroblastomas and liver tumors (Pajer, unpublished) provided negative results.

4.8 The Future of Models Based on Chicken Retroviruses

During the last years the effort to identify genes implicated in cancer induction and development ("cancer genes") has intensified. High throughput forward genetic screens based on insertional mutagenesis have been employed and transposons as a new versatile insertional mutagens have been introduced [37, 150]. Supremely informative data started to be acquired by genome-wide analyses of human tumors [29, 144]. All this effort is driven by hopes that with detailed knowledge of "genetic landscape of cancer" it will be possible to design new therapeutics targeted directly against the crucial mutated cancer genes. Such therapeutics are expected to be more efficient and have less detrimental side effects compared to the conventional ones that are directed nondiscriminatingly against all rapidly dividing cells.

Accumulating results show that the number of genes that can be implicated even in one particular type of cancer is formidable albeit in a single tumor only several of them are mutated. Thus, except for a few commonest oncogenes and tumor suppressor genes, each potential cancer gene is mutated only in a very small proportion of the tumors. To be able to design personalized therapy in every cancer case enormous numbers of targeted therapeutics would have to be developed. Fortunately, majority of the discovered cancer genes fall into a handful of signaling pathways which are, to a large extent, shared between different tumor classes. With little exceptions only one member of the pathway is mutated in one tumor reaffirming the notion that deregulation of a particular pathways, not of a particular genes, is what matters in tumorigenesis. Consequently, it should suffice to develop therapeutics targeted against the selected key components of the pathways, not against all their members.

The genome-wide analyses of human tumors must comprise huge sample collections due to large heterogeneity of tumors, large number of alternative cancer genes and presence, in each tumor, of immense numbers of fortuitous passenger mutations not related to cancerogenesis. The results have to be processed using statistical analyses and the screens provide rather suspect than undoubted cancer genes [162]. Moreover, these analyses reveal only static endpoint-set of genetic alterations not showing the history and dynamics of tumor development; also the fundamental importance of stromal cells and tumor microenvironment is missed out. All what was said underscores the imperative that extensive cancer gene screens were followed by experiments—to validate the gene's causative role in cancerogenesis, show in what step they act, establish their grouping into individual signaling pathways, elucidate their mutual cooperation and study the role of microenvironment. Since all this is impossible to pursue clinically, animal models are indispensable.

While mouse models of human cancerogenesis are now generally accepted [44] a question remains if and to what extent the models based on evolutionary more distant species like chickens are relevant to humans. The answer is certainly positive as majority of organ systems are shared between chickens, humans and mice, their tissues have the same organisational principles and contain the same types of intercommunicating cells. The signal transduction pathways including those involved in tumorigenesis are conserved over much larger evolutionary distances, beyond the phylum chordata. Nevertheless, different elements of the pathway or even different pathways are often preferentially mutated in different species during oncogenic transformation of a particular cell type. Such predisposition suggests that differences in regulatory networks between different vertebrate species do exist or, at least, that in some pathways different elements in different species are the rate-limiting ones. Well known and frequently encountered difference is a varying pattern of tissue specific expression of functionally redundant genes in different species. An illustrative example is induction of retinoblastoma in humans versus mice. In humans, retinoblastoma originates after *Rb1* gene has been homozygously inactivated in a single cell of developing retina. To induce retinoblastoma in mice, both *Rb1* and its relative *p107* must be inactivated. Different sensitivity to *Rb1* inactivation results from the simple fact that both *Rb1* and *p107* are expressed in developing mouse retina and can compensate for each other's inactivation while only *Rb1* is expressed during retinal development in humans [36].

As a matter of fact, certain cancerogenesis-influencing differences in genetic background exist even between strains of the same species and between individuals from non-inbred population, which includes human population as well. All the more significant differences must be expected between members of different animal classes like between humans and chickens. Faithful modeling of a specific human cancer is challenging even in mouse system [44, 125], therefore chickens do not represent a suitable system for this type of modeling. When we use the term "animal model" in the next paragraphs, we mean nothing more than productive and convenient instruments to search for cancer genes and to study their function and cooperation.

Cancer gene screens in animals, especially those based on mouse retroviruses, have been very successful in gathering data applicable for outlining genetic landscape of cancer. For a long time, however, the investigation has been limited to hematopoetic and mammary tumors due to a limited tissue tropism of the used mouse retroviruses (MLV and MMTV). To progress further it was essential to extend the range of analyzed tumor types and, ideally, to create models corresponding to the most relevant human tumors. This objective was fulfilled with great success by the development of universal transposon-based technology [26, 73, 141].

The tropism of avian retroviruses is not as strictly limited as is the case for MLV and MMTV. In addition to neoplasias described above (B-lymphoma, erythroblastosis, nephroblastoma, lung sarcoma and hepatocarcinoma) simple avian retroviruses are capable of inducing many other tumor types, e.g. myeloid leukosis, histiocytic sarcoma, angiosarcoma, fibrosarcoma, renal carcinoma, mesothelioma, and pancreatic adenocarcinoma [15, 110, 113, 121, 137, 153]. Though the penetrance of these malignancies is mostly very low, the fundamental ability to transform a wide range of target cells is apparent. To carry out cancer gene screens, however, the frequency of rare tumor classes have to be increased. Below we suggest two ways how to achieve that; no doubt further possibilities can be conceived or will come up over time.

Nongenetic non-cell-autonomous factors that influence tissue homeostasis, such as wound healing and local inflammation, have long been known to play an important role in both experimental and human oncogenesis [69, 74, 86, 154]. The industasis phenomenon (Section 4.6) illustrates that these factors may as well operate during oncogenesis through insertional mutagenesis. Classical tumor promoters like phorbol esters fall into the same sort of factors. Many different tumor promoters have been described, some of them tissue specific [87]. Hence, the penetrance of the rare retrovirus-induced cancer classes can be increased by tumor promoters either through their tissue-specific delivery or by using the tissue-specific ones. The important point is that assistance of tumor promoters does not bring any additional genetic changes into the cells and that tumors originate from the same pool of dormant mutagenized cells that give rise to the rare tumors when no promotion is applied (see Section 4.6). In addition to the increased penetrance, tumor promoters could also reduce latency thus facilitating study of tumor progression.

The oncogenic spectrum of simple retroviruses is primarily determined by the promoter/enhancer sequences in provirus LTR—see, for example, the properties of MAV/RAV recombinants [67]. Consistently, enhancer mutations have been shown to change oncogenic spectrum of the retrovirus [39, 139]. This is not surprising since tissue-specific activity of promoter/enhancer not only determines the tissue where the virus will replicate best, but also the cell type where the integrated provirus will affect neighboring genes most vigorously. Manipulating the enhancer is thus another way how to shift retroviral oncogenic spectrum. For example, segments of known tissue specific promoters/enhancers may be inserted into viral U3 region. Definitely, problems with the proper enhancer design can be anticipated, resulting from our lack of understanding how combined enhancer segments interact with each other. These problems, however, may be surmounted by generating many alternative variants (e.g. by shotgun approach when assembling the LTR or by employing random mutagenesis) and selecting the successful variants in cell culture or animals. Similar approaches have already been utilized, though with varying success, see e.g. [5, 42, 43, 52, 88].

To understand fully the relevance of results obtained in animal cancer gene screens, we must contemplate also another factor, independent of evolutionary distance, responsible for the incomplete overlap of cancer genes mutated in experimental model versus human cancer. It is the dissimilar molecular mechanisms of cancer gene mutation/deregulation. In humans the cancer genes are affected largely by point mutations and chromosome rearrangements while in animal screens they are altered by provirus/transposon insertions. This introduces bias to the spectrum of mutated cancer genes because (i) oncogenic changes of a cancer gene introduced by point mutations or chromosome rearrangement cannot always be accurately simulated by proviral insertion and (ii) the starting distributions (prior to oncogenic selection) of chromosome rearrangements, point mutations and proviral insertions, respectively, throughout the genome are uneven and differ from each other.

(i) Evidently, the effect of human cancer gene amplification (with resulting overexpression) or, conversely, the effect of gene knockout/deactivation can be reproduced by provirus/transposon insertion. Also the effect of chromosomal translocation can mostly be simulated by insertional mutagenesis, see c–myc activation by translocation in human Burkitt lymphomas [32] as well as by insertion in chicken B-lymphomas or mouse T-lymphomas [30, 126]. In these cases the full-length cancer gene is overexpressed as a result of being placed under the control of strong non-authentic promoter/enhancer. Frequently, however, the translocated human cancer gene is not only overexpressed, but also truncated and fused to another gene which results in production of a chimeric protein [94]. Effects of such translocations can be mimicked by insertional mutagenesis only in those instances when the overexpression and/or truncation alone, without a fusion, is sufficient to activate the gene's oncogenic potential or when the cancer gene's fusion partner can be substituted by retrovirus/transposon-derived sequences like in the case of *bcr–abl* and *gag–abl* fusions [161]. Point mutations, the typical mode of oncogenic activation of many cancer genes in humans, cannot, obviously, be inflicted by provirus/transposon insertion. In spite of it, many of these genes repeatedly emerge in insertional mutagenesis screens. It is so because they can also be activated by other mechanisms than point mutation, e.g. by amplification/overexpression as exemplified by *Ha–ras* gene [83, 108, 120, 145, 148] or by truncation as exemplified by *B–raf* gene [26, 90]. Certainly, cancer genes mutated by different mechanisms may have not fully equivalent oncogenic properties; that, however, is irrelevant since we are discussing forward genetic screening, not modeling of a particular human tumor.

(ii) Even if particular gene can be, in principle, oncogenically activated by different mechanisms (for example by chromosomal translocation in humans and by provirus/transposon insertion in experimental animals) the relative frequency of such events may be very different. The primary choice of chromosomal recombination sites as well as selection of integration sites are far from random. Particularly the frequency of breakage and recombination in different parts of human genome is highly non-uniform as manifested by the existence of hot spots of recombination and chromosomal fragile sites. This bias is caused by specific local DNA and chromatin features that are unrelated to function of the residing genes. Different features are relevant for chromosomal recombination versus provirus/transposon insertion [17, 18, 166, 169]. In consequence, when particular pathway is to be deregulated, different elements of the pathway will be preferentially mutated in humans versus experimental animals.

The propensity to mutate preferentially different genes exists even between different insertional mutagens. For instance, *H–ras* gene is frequently activated by retroviral proviruses [83, 108, 145, 148] but never by Sleeping Beauty transposon [141]. Even closely related retroviruses that induce identical subclass of tumors but differ in their enhancer sequences may activate very different sets of genes [65]. Most plausible explanation of this phenomenon is that there exists significant (yet so far overlooked) interaction of inserted enhancer sequences with the host regulatory sequences at the site of insertion; this interaction may be productive in case of one enhancer and counterproductive in case of another one.

There are some practical advantages of chicken system worth to mention. For example manipulating and monitoring chicken embryo is much more feasible and convenient. A great advantage of chicken system is that gene activation is preferentially carried out by the promoter insertion mechanism, which is characterized by a narrow region of insertions in front of the activated gene. This eliminates dilemma frequently encountered in mouse models: which of the genes in the CIS locus is the cancer-related one, see [53, 131].

One disadvantage of chicken models must also be mentioned here. It is the absence of the huge variety of strains predisposed to specific tumor types due to the presence of transgenic or knockout cancer gene alleles, which are available in mouse system. Insertional mutagenesis in genetically engineered animals is a powerful method for studying cancer genes cooperation. The first generation of genetically engineered mice had, however, certain week sides: the oncogenic mutation was expressed already during embryogenesis and in all cells of the target tissue, which is very different condition compare to spontaneous tumor origination and may significantly influence course and outcome of tumorigenesis [44, 130, 171]. To preclude this drawback, new generations of genetically engineered mice are being constructed in which the oncogene/tumor suppressor gene is activated/deleted later during the lifetime and only in sporadic cells [66, 89, 95, 157].

The technology for generating transgenic chickens is now at hand [51, 70]; however, the range of genetically modified chicken strains can never come close to the variety available in mouse system. Fortunately, production of predisposed genetically modified chickens can be conveniently substituted by using in vivo gene transfer mediated by retroviral or transposon-based vectors that carry an activated oncogene, inhibitory RNA directed against tumor suppressor gene or dominant-negative form of tumor suppressor gene. This technique, moreover, does not comprise the drawbacks of the transgenic approach mentioned above. Combination of insertional mutagenesis and in vivo gene transfer would enable not only to investigate cancer genes cooperation in chicken models but also to redirect the retrovirus tumorigenic activity into selected tissue either by tissue-specific vector delivery or through the tissue-specific capacity of the delivered cancer gene. Efficacy of in vivo gene transfer strategies has already been proven in mouse system [57, 64, 73, 78].

The conclusion of the above discussion is that no cancer gene screening model fully reproduces corresponding human cancer. Multiple models including multiple model animals must be combined to cover all players involved in the disease. Comparing and integrating results obtained in different models and in genome-wide screening of human tumors may prove to be especially beneficial for delineating the genetic landscape of cancer. Thus, we are convinced that models based on chicken retroviruses can still considerably contribute to our knowledge of genes, pathways and networks implicated in cancerogenesis.

References

1. Adhya, S., & Gottesman, M. (1982). Promoter occlusion: Transcription through a promoter may inhibit its activity. *Cell, 29*, 939–944.
2. Ansieau, S., Bastid, J., Doreau, A., Morel, A. P., Bouchet, B. P., Thomas, C., et al. (2008). Induction of EMT by twist proteins as a collateral effect of tumor-promoting inactivation of premature senescence. *Cancer Cell, 14*, 79–89.
3. Arrigo, S., Yun, M., & Beemon, K. (1987). cis-acting regulatory elements within gag genes of avian retroviruses. *Molecular and Cellular Biology, 7*, 388–397.
4. Baba, T. W., & Humphries, E. H. (1985). Formation of a transformed follicle is necessary but not sufficient for development of an avian leukosis virus-induced lymphoma. *Proceedings of the National Academy of Sciences of the USA, 82*, 213–216.
5. Bahrami, S., Duch, M., & Pedersen, F. S. (2004). Change of tropism of SL3-2 murine leukemia virus, using random mutational libraries. *Journal of Virology, 78*, 9343–9351.

6. Ball, J. K., Diggelmann, H., Dekaban, G. A., Grossi, G. F., Semmler, R., Waight, P. A., et al. (1988). Alterations in the U3 region of the long terminal repeat of an infectious thymotropic type B retrovirus. *Journal of Virology*, *62*, 2985–2993.
7. Beard, J. W., Chabot, J. F., Beard, D., Heine, U., & Houts, G. E. (1976). Renal neoplastic response to leukosis virus strains BAI A (avian myeloblastosis virus) and MC29. *Cancer Research*, *36*, 339–353.
8. Beemon, K. L., O'Reilly, M. M., Smith, M. R., Smith, R. E., Dunkel, I., & Hayward, W. S. (1997). Rapid induction of B-cell lymphomas by avian leukosis virus. *Leukemia*, *11*(Suppl 3), 179–182.
9. Bird, K. J., Semus, H. L., & Ruddell, A. (1999). Resistance to avian leukosis virus lymphomagenesis occurs subsequent to proviral c-myc integration. *Oncogene*, *18*, 201–209.
10. Boerkoel, C. F., & Kung, H. J. (1992). Transcriptional interaction between retroviral long terminal repeats (LTRs): Mechanism of 5′ LTR suppression and 3′ LTR promoter activation of c-myc in avian B-cell lymphomas. *Journal of Virology*, *66*, 4814–4823.
11. Boni-Schnetzler, M., Boni, J., Ferdinand, F. J., & Franklin, R. M. (1985). Developmental and molecular aspects of nephroblastomas induced by avian myeloblastosis-associated virus 2-O. *Journal of Virology*, *55*, 213–222.
12. Brandvold, K. A., Ewert, D. L., Kent, S. C., Neiman, P., & Ruddell, A. (2001). Blocked B cell differentiation and emigration support the early growth of Myc-induced lymphomas. *Oncogene*, *20*, 3226–3234.
13. Brauer, P. M., & Tyner, A. L. (2009). RAKing in AKT: A tumor suppressor function for the intracellular tyrosine kinase FRK. *Cell Cycle*, *8*, 2728–2732.
14. Broussard, D. R., Mertz, J. A., Lozano, M., & Dudley, J. P. (2002). Selection for c-myc integration sites in polyclonal T-cell lymphomas. *Journal of Virology*, *76*, 2087–2099.
15. Brown, D. W., Blais, B. P., & Robinson, H. L. (1988). Long terminal repeat (LTR) sequences, env, and a region near the 5′ LTR influence the pathogenic potential of recombinants between Rous-associated virus types 0 and 1. *Journal of Virology*, *62*, 3431–3437.
16. Burmester, B. R., & Waters, N. F. (1955). The role of the infected egg in the transmission of visceral lymphomatosis. *Poultry Science*, *34*, 1415–1429.
17. Bushman, F., Lewinski, M., Ciuffi, A., Barr, S., Leipzig, J., Hannenhalli, S., et al. (2005). Genome-wide analysis of retroviral DNA integration. *Nature Reviews Microbiology*, *3*, 848–858.
18. Buttel, I., Fechter, A., & Schwab, M. (2004). Common fragile sites and cancer: Targeted cloning by insertional mutagenesis. *Annals of the New York Academy Science*, *1028*, 14–27.
19. Cance, W. G., Craven, R. J., Bergman, M., Xu, L., Alitalo, K., & Liu, E. T. (1994). Rak, a novel nuclear tyrosine kinase expressed in epithelial cells. *Cell Growth and Differentiation*, *5*, 1347–1355.
20. Ceci, J. D., Patriotis, C. P., Tsatsanis, C., Makris, A. M., Kovatch, R., Swing, D. A., et al. (1997). Tpl-2 is an oncogenic kinase that is activated by carboxy-terminal truncation. *Genes and Development*, *11*, 688–700.
21. Chandrasekharan, S., Qiu, T. H., Alkharouf, N., Brantley, K., Mitchell, J. B., & Liu, E. T. (2002). Characterization of mice deficient in the Src family nonreceptor tyrosine kinase Frk/rak. *Molecular and Cellular Biology*, *22*, 5235–5247.
22. Cheng, G. Z., Zhang, W., & Wang, L. H. (2008). Regulation of cancer cell survival, migration, and invasion by Twist: AKT2 comes to interplay. *Cancer Research*, *68*, 957–960.
23. Clurman, B. E., & Hayward, W. S. (1989). Multiple proto-oncogene activations in avian leukosis virus-induced lymphomas: Evidence for stage-specific events. *Molecular and Cellular Biology*, *9*, 2657–2664.
24. Cochrane, A. W., McNally, M. T., & Mouland, A. J. (2006). The retrovirus RNA trafficking granule: From birth to maturity. *Retrovirology*, *3*, 18.
25. Coffin, J. M., Hughes, S. H., & Varmus, H. E. (1997). *Retroviruses*. New York: Cold Spring Harbor Laboratory Press.

26. Collier, L. S., Carlson, C. M., Ravimohan, S., Dupuy, A. J., & Largaespada, D. A. (2005). Cancer gene discovery in solid tumours using transposon-based somatic mutagenesis in the mouse. *Nature, 436*, 272–276.
27. Cooper, G. M., & Neiman, P. E. (1981). Two distinct candidate transforming genes of lymphoid leukosis virus-induced neoplasms. *Nature, 292*, 857–8.
28. Cooper, M. D., Payne, L. N., Dent, P. B., Burmester, B. R., & Good, R. A. (1968). Pathogenesis of avian lymphoid leukosis. I. Histogenesis. *Journal of the National Cancer Institute, 41*, 373–378.
29. Copeland, N. G., & Jenkins, N. A. (2009). Deciphering the genetic landscape of cancer - from genes to pathways. *Trends in Genetics, 25*, 455–462.
30. Corcoran, L. M., Adams, J. M., Dunn, A. R., & Cory, S. (1984). Murine T lymphomas in which the cellular myc oncogene has been activated by retroviral insertion. *Cell, 37*, 113–122.
31. Crittenden, L. B., Hayward, W. S., Hanafusa, H., & Fadly, A. M. (1980). Induction of neoplasms by subgroup E recombinants of exogenous and endogenous avian retroviruses (Rous-associated virus type 60). *Journal of Virology, 33*, 915–919.
32. Croce, C. M., & Nowell, P. C. (1985). Molecular basis of human B cell neoplasia. *Blood, 65*, 1–7.
33. Cullen, B. R., Lomedico, P. T., & Ju, G. (1984). Transcriptional interference in avian retroviruses–implications for the promoter insertion model of leukaemogenesis. *Nature, 307*, 241–245.
34. Dabrowska, M. J., Dybkaer, K., Johnsen, H. E., Wang, B., Wabl, M., & Pedersen, F. S. (2009). Loss of MicroRNA targets in the 3′ untranslated region as a mechanism of retroviral insertional activation of growth factor independence 1. *Journal of Virology, 83*, 8051–8061.
35. de Laat, W., & Grosveld, F. (2003). Spatial organization of gene expression: The active chromatin hub. *Chromosome Research, 11*, 447–459.
36. Donovan, S. L., Schweers, B., Martins, R., Johnson, D., & Dyer, M. A. (2006). Compensation by tumor suppressor genes during retinal development in mice and humans. *BMC Biology, 4*, 14.
37. Dupuy, A. J., Jenkins, N. A., & Copeland, N. G. (2006). Sleeping beauty: A novel cancer gene discovery tool. *Human Molecular Genetics, 15*(Spec No 1), R75–R79.
38. Eis, P. S., Tam, W., Sun, L., Chadburn, A., Li, Z., Gomez, M. F., et al. (2005). Accumulation of miR-155 and BIC RNA in human B cell lymphomas. *Proceedings of the National Academy of Sciences of the USA, 102*, 3627–3632.
39. Ejegod, D., Sorensen, K. D., Mossbrugger, I., Quintanilla-Martinez, L., Schmidt, J., & Pedersen, F. S. (2009). Control of pathogenicity and disease specificity of a T-lymphomagenic gammaretrovirus by E-box motifs but not by an overlapping glucocorticoid response element. *Journal of Virology, 83*, 336–346.
40. El Kaderi, B., Medler, S., Raghunayakula, S., & Ansari, A. (2009). Gene looping is conferred by activator-dependent interaction of transcription initiation and termination machineries. *Journal of Biological Chemistry, 284*, 25015–25025.
41. Ellerman, V., & Bang, O. (1908). Experimentelle Leukämie bei Hühnern. *Zentralblatt fur Bakteriologi, Parasitenkunde, Infektionskrinkheiten und Hygiene Abteilung Originale, 46*, 595–609.
42. Ethelberg, S., Sorensen, A. B., Schmidt, J., Luz, A., & Pedersen, F. S. (1997). An SL3-3 murine leukemia virus enhancer variant more pathogenic than the wild type obtained by assisted molecular evolution in vivo. *Journal of Virology, 71*, 9796–9799.
43. Feuer, G., & Fan, H. (1990). Substitution of murine transthyretin (prealbumin) regulatory sequences into the Moloney murine leukemia virus long terminal repeat yields infectious virus with altered biological properties. *Journal of Virology, 64*, 6130–6140.

44. Frese, K. K., & Tuveson, D. A. (2007). Maximizing mouse cancer models. *Nature Reviews Cancer*, *7*, 645–658.
45. Fung, Y. K., Fadly, A. M., Crittenden, L. B., & Kung, H. J. (1981). On the mechanism of retrovirus-induced avian lymphoid leukosis: Deletion and integration of the proviruses. *Proceedings of the National Academy of Sciences of the USA*, *78*, 3418–3422.
46. Fung, Y. K., Lewis, W. G., Crittenden, L. B., & Kung, H. J. (1983). Activation of the cellular oncogene c-erbB by LTR insertion: Molecular basis for induction of erythroblastosis by avian leukosis virus. *Cell*, *33*, 357–368.
47. Gentile, A., Trusolino, L., & Comoglio, P. M. (2008). The Met tyrosine kinase receptor in development and cancer. *Cancer Metastasis Reviews*, *27*, 85–94.
48. Gong, M., Semus, H. L., Bird, K. J., Stramer, B. J., & Ruddell, A. (1998). Differential selection of cells with proviral c-myc and c-erbB integrations after avian leukosis virus infection. *Journal of Virology*, *72*, 5517–5525.
49. Goodenow, M. M., & Hayward, W. S. (1987). 5′ long terminal repeats of myc-associated proviruses appear structurally intact but are functionally impaired in tumors induced by avian leukosis viruses. *Journal of Virology*, *61*, 2489–2498.
50. Goodwin, R. G., Rottman, F. M., Callaghan, T., Kung, H. J., Maroney, P. A., & Nilsen, T. W. (1986). c-erbB activation in avian leukosis virus-induced erythroblastosis: Multiple epidermal growth factor receptor mRNAs are generated by alternative RNA processing. *Molecular and Cellular Biology*, *6*, 3128–3133.
51. Han, J. Y. (2009). Germ cells and transgenesis in chickens. *Comparative Immunology, Microbiology and Infectious Diseases*, *32*, 61–80.
52. Hanecak, R., Pattengale, P. K., & Fan, H. (1991). Deletion of a GC-rich region flanking the enhancer element within the long terminal repeat sequences alters the disease specificity of Moloney murine leukemia virus. *Journal of Virology*, *65*, 5357–5363.
53. Hanlon, L., Barr, N. I., Blyth, K., Stewart, M., Haviernik, P., Wolff, L., et al. (2003). Long-range effects of retroviral insertion on c-myb: Overexpression may be obscured by silencing during tumor growth in vitro. *Journal of Virology*, *77*, 1059–1068.
54. Hayward, W. S., Neel, B. G., & Astrin, S. M. (1981a). Activation of a cellular onc gene by promoter insertion in ALV-induced lymphoid leukosis. *Nature*, *290*, 475–480.
55. Hayward, W. S., Neel, B. G., Fang, J., Robinson, H. L., & Astrin, S. M. (1981b). Avian lymphoid leukosis is correlated with the appearance of discrete new RNAs containing viral and cellular genetic information. *Hematology and Blood Transfusion*, *26*, 439–444.
56. Heine, U., De The, G., Ishiguro, H., Sommer, J. R., Beard, D., & Beard, J. W. (1962). Multiplicity of cell response to the BAI strain A (myeloblastosis) avian tumor virus. II. Nephroblastoma (Wilms' tumor): Ultrastructure. *Journal of the National Cancer Institute*, *29*, 41–105.
57. Hemann, M. T., Fridman, J. S., Zilfou, J. T., Hernando, E., Paddison, P. J., Cordon-Cardo, C., et al. (2003). An epi-allelic series of p53 hypomorphs created by stable RNAi produces distinct tumor phenotypes in vivo. *Nature Genetics*, *33*, 396–400.
58. Herman, S. A., & Coffin, J. M. (1986). Differential transcription from the long terminal repeats of integrated avian leukosis virus DNA. *Journal of Virology*, *60*, 497–505.
59. Herman, S. A., & Coffin, J. M. (1987). Efficient packaging of readthrough RNA in ALV: Implications for oncogene transduction. *Science*, *236*, 845–848.
60. Hosoya, N., Qiao, Y., Hangaishi, A., Wang, L., Nannya, Y., Sanada, M., et al. (2005). Identification of a SRC-like tyrosine kinase gene, FRK, fused with ETV6 in a patient with acute myelogenous leukemia carrying a t(6;12)(q21;p13) translocation. *Genes Chromosomes and Cancer*, *42*, 269–279.
61. Huppi, K., Siwarski, D., Shaughnessy, J. D., Jr., & Mushinski, J. F. (1993). Co-amplification of c-myc/pvt-1 in immortalized mouse B-lymphocytic cell lines results in a novel pvt-1/AJ-1 transcript. *International Journal of Cancer*, *53*, 493–498.
62. Ishiguro, H., Beard, D., Sommer, J. R., Heine, U., De The, G., & Beard, J. W. (1962). Multiplicity of cell response to the BAI strain A (myeloblastosis) avian tumor virus. II.

Nephroblastoma (Wilms' tumor): Gross and microscopic pathology. *Journal of the National Cancer Institute*, *29*, 1–17.
63. Jiang, W., Kanter, M. R., Dunkel, I., Ramsay, R. G., Beemon, K. L., & Hayward, W. S. (1997). Minimal truncation of the c-myb gene product in rapid-onset B-cell lymphoma. *Journal of Virology*, *71*, 6526–6533.
64. Johansson, F. K., Brodd, J., Eklof, C., Ferletta, M., Hesselager, G., Tiger, C. F., et al. (2004). Identification of candidate cancer-causing genes in mouse brain tumors by retroviral tagging. *Proceedings of the National Academy of Sciences of the USA*, *101*, 11334–11337.
65. Johnson, C., Lobelle-Rich, P. A., Puetter, A., & Levy, L. S. (2005). Substitution of feline leukemia virus long terminal repeat sequences into murine leukemia virus alters the pattern of insertional activation and identifies new common insertion sites. *Journal of Virology*, *79*, 57–66.
66. Johnson, L., Mercer, K., Greenbaum, D., Bronson, R. T., Crowley, D., Tuveson, D. A., et al. (2001). Somatic activation of the K-ras oncogene causes early onset lung cancer in mice. *Nature*, *410*, 1111–1116.
67. Joliot, V., Khelifi, C., Wyers, M., Dambrine, G., Lasserre, F., Lemercier, P., et al. (1996). The noncoding and surface envelope coding sequences of myeloblastosis-associated virus are respectively responsible for nephroblastoma development and renal hyperplasia. *Journal of Virology*, *70*, 2576–2580.
68. Joliot, V., Martinerie, C., Dambrine, G., Plassiart, G., Brisac, M., Crochet, J., et al. (1992). Proviral rearrangements and overexpression of a new cellular gene (nov) in myeloblastosis-associated virus type 1-induced nephroblastomas. *Molecular and Cellular Biology*, *12*, 10–21.
69. Jones, F. S., & Rous, P. (1914). On the cause of the localization of secondary tumors at points of injury. *J Exp Med*, *20*, 404–412.
70. Kalina, J., Senigl, F., Micakova, A., Mucksova, J., Blazkova, J., Yan, H., et al. (2007). Retrovirus-mediated in vitro gene transfer into chicken male germ line cells. *Reproduction*, *134*, 445–453.
71. Kanter, M. R., Smith, R. E., & Hayward, W. S. (1988). Rapid induction of B-cell lymphomas: Insertional activation of c-myb by avian leukosis virus. *Journal of Virology*, *62*, 1423–1432.
72. Kawai, S., & Yamamoto, T. (1970). Isolation of different kinds of non-virus producing chick cells transformed by Schmidt-Ruppin strain (subgroup A) of Rous sarcoma virus. *Japanese Journal of Experimental Medicine*, *40*, 243–256.
73. Keng, V. W., Villanueva, A., Chiang, D. Y., Dupuy, A. J., Ryan, B. J., Matise, I., et al. (2009). A conditional transposon-based insertional mutagenesis screen for genes associated with mouse hepatocellular carcinoma. *Nature Biotechnology*, *27*, 264–274.
74. Kluwe, J., Mencin, A., & Schwabe, R. F. (2009). Toll-like receptors, wound healing, and carcinogenesis. *Journal of Molecular Medicine*, *87*, 125–138.
75. Kool, J., & Berns, A. (2009). High-throughput insertional mutagenesis screens in mice to identify oncogenic networks. *Nature Reviews Cancer*, *9*, 389–399.
76. Koon, H. B., Ippolito, G. C., Banham, A. H., & Tucker, P. W. (2007). FOXP1: A potential therapeutic target in cancer. *Expert Opinion on Therapeutic Targets*, *11*, 955–965.
77. Lax, I., Kris, R., Sasson, I., Ullrich, A., Hayman, M. J., Beug, H., et al. (1985). Activation of c-erbB in avian leukosis virus-induced erythroblastosis leads to the expression of a truncated EGF receptor kinase. *EMBO Journal*, *4*, 3179–3182.
78. Lewis, B. C., Klimstra, D. S., Socci, N. D., Xu, S., Koutcher, J. A., & Varmus, H. E. (2005). The absence of p53 promotes metastasis in a novel somatic mouse model for hepatocellular carcinoma. *Molecular and Cellular Biology*, *25*, 1228–1237.
79. Li, T., & Zhang, J. (2002). Intramolecular recombinations of Moloney murine leukemia virus occur during minus-strand DNA synthesis. *Journal of Virology*, *76*, 9614–9623.
80. Lin, Q., Chen, Q., Lin, L., Smith, S., & Zhou, J. (2007). Promoter targeting sequence mediates enhancer interference in the Drosophila embryo. *Proceedings of the National Academy of Sciences of the USA*, *104*, 3237–3242.

81. Linial, M., & Groudine, M. (1985). Transcription of three c-myc exons is enhanced in chicken bursal lymphoma cell lines. *Proceedings of the National Academy of Sciences of the USA*, *82*, 53–57.
82. Lu, Y., Yao, H. P., & Wang, M. H. (2007). Multiple variants of the RON receptor tyrosine kinase: Biochemical properties, tumorigenic activities, and potential drug targets. *Cancer Letters*, *257*, 157–164.
83. Lund, A. H., Turner, G., Trubetskoy, A., Verhoeven, E., Wientjens, E., Hulsman, D., et al. (2002). Genome-wide retroviral insertional tagging of genes involved in cancer in Cdkn2a-deficient mice. *Nature Genetics*, *32*, 160–165.
84. Maciolek, N. L., & McNally, M. T. (2007). Serine/arginine-rich proteins contribute to negative regulator of splicing element-stimulated polyadenylation in rous sarcoma virus. *Journal of Virology*, *81*, 11208–11217.
85. Maihle, N. J., Raines, M. A., Flickinger, T. W., & Kung, H. J. (1988). Proviral insertional activation of c-erbB: Differential processing of the protein products arising from two alternate transcripts. *Molecular and Cellular Biology*, *8*, 4868–4876.
86. Mantovani, A., Allavena, P., Sica, A., & Balkwill, F. (2008). Cancer-related inflammation. *Nature*, *454*, 436–444.
87. Marks, F., Schwartz, M., & Furstenberg, G. (1995). Promotion and cocarcinogenesis. In J. C. Arcos (ed.), *Chemical induction of cancer: Modulation and combination effects. An inventory of the many factors which influence carcinogenesis* (pp. 123–184.). Boston: Burkhauser.
88. McGee-Estrada, K., Palmarini, M., Hallwirth, C., & Fan, H. (2005). A Moloney murine leukemia virus driven by the Jaagsiekte sheep retrovirus enhancers shows enhanced specificity for infectivity in lung epithelial cells. *Virus Genes*, *31*, 257–263.
89. Meuwissen, R., Linn, S. C., van der Valk, M., Mooi, W. J., & Berns, A. (2001). Mouse model for lung tumorigenesis through Cre/lox controlled sporadic activation of the K-Ras oncogene. *Oncogene*, *20*, 6551–6558.
90. Michaloglou, C., Vredeveld, L. C., Mooi, W. J., & Peeper, D. S. (2008). BRAF(E600) in benign and malignant human tumours. *Oncogene*, *27*, 877–895.
91. Mikkers, H., Allen, J., Knipscheer, P., Romeijn, L., Hart, A., Vink, E., et al. (2002). High-throughput retroviral tagging to identify components of specific signaling pathways in cancer. *Nature Genetics*, *32*, 153–159.
92. Miles, B. D., & Robinson, H. L. (1985). High-frequency transduction of c-erbB in avian leukosis virus-induced erythroblastosis. *Journal of Virology*, *54*, 295–303.
93. Miller, J. T., & Stoltzfus, C. M. (1992). Two distant upstream regions containing cis-acting signals regulating splicing facilitate 3′-end processing of avian sarcoma virus RNA. *Journal of Virology*, *66*, 4242–4251.
94. Mitelman, F., Johansson, B., & Mertens, F. (2007). The impact of translocations and gene fusions on cancer causation. *Nature Reviews Cancer*, *7*, 233–245.
95. Muzumdar, M. D., Luo, L., & Zong, H. (2007). Modeling sporadic loss of heterozygosity in mice by using mosaic analysis with double markers (MADM. *Proceedings of the National Academy of Sciences of the USA*, *104*, 4495–4500.
96. Neel, B. G., Hayward, W. S., Robinson, H. L., Fang, J., & Astrin, S. M. (1981). Avian leukosis virus-induced tumors have common proviral integration sites and synthesize discrete new RNAs: Oncogenesis by promoter insertion. *Cell*, *23*, 323–334.
97. Neil, J. C., Fulton, R., Rigby, M., & Stewart, M. (1991). Feline leukaemia virus: Generation of pathogenic and oncogenic variants. *Current Topics in Microbiology and Immunology*, *171*, 67–93.
98. Neiman, P. E., Grbic, J. J., Polony, T. S., Kimmel, R., Bowers, S. J., Delrow, J., et al. (2003). Functional genomic analysis reveals distinct neoplastic phenotypes associated with c-myb mutation in the bursa of fabricius. *Oncogene*, *22*, 1073–1086.
99. Nilsen, T. W., Maroney, P. A., Goodwin, R. G., Rottman, F. M., Crittenden, L. B., Raines, M. A., et al. (1985). c-erbB activation in ALV-induced erythroblastosis: Novel RNA processing

and promoter insertion result in expression of an amino-truncated EGF receptor. *Cell, 41*, 719–726.
100. Noori-Daloii, M. R., Swift, R. A., Kung, H. J., Crittenden, L. B., & Witter, R. L. (1981). Specific integration of REV proviruses in avian bursal lymphomas. *Nature, 294*, 574–576.
101. Norton, P. A., & Coffin, J. M. (1987). Characterization of Rous sarcoma virus sequences essential for viral gene expression. *Journal of Virology, 61*, 1171–1179.
102. Nusse, R., Theunissen, H., Wagenaar, E., Rijsewijk, F., Gennissen, A., Otte, A., et al. (1990). The Wnt-1 (int-1) oncogene promoter and its mechanism of activation by insertion of proviral DNA of the mouse mammary tumor virus. *Molecular and Cellular Biology, 10*, 4170–4179.
103. Olah, I., & Glick, B. (1978). The number and size of the follicular epithelium (FE) and follicles in the bursa of Fabricius. *Poultry Science, 57*, 1445–1450.
104. Olayioye, M. A., Neve, R. M., Lane, H. A., & Hynes, N. E. (2000). The ErbB signaling network: Receptor heterodimerization in development and cancer. *EMBO Journal, 19*, 3159–3167.
105. O'Sullivan, C. T., Polony, T. S., Paca, R. E., & Beemon, K. L. (2002). Rous sarcoma virus negative regulator of splicing selectively suppresses SRC mRNA splicing and promotes polyadenylation. *Virology, 302*, 405–412.
106. Pajer, P., Karafiat, V., Pecenka, V., Prukova, D., Dudlova, J., Plachy, J., et al. (2009). Industasis, a promotion of tumor formation by nontumorigenic stray cells. *Cancer Research, 69*, 4605–4612.
107. Pajer, P., Pecenka, V., Karafiat, V., Kralova, J., Horejsi, Z., & Dvorak, M. (2003). The twist gene is a common target of retroviral integration and transcriptional deregulation in experimental nephroblastoma. *Oncogene, 22*, 665–673.
108. Pajer, P., Pecenka, V., Kralova, J., Karafiat, V., Prukova, D., Zemanova, Z., et al. (2006). Identification of potential human oncogenes by mapping the common viral integration sites in avian nephroblastoma. *Cancer Research, 66*, 78–86.
109. Parghi, S. S., Brandvold, K. A., Bowers, S. J., Neiman, P. E., & Ruddell, A. (2004). Reduced Myc overexpression and normal B-cell differentiation mediate resistance to avian leukosis virus lymphomagenesis. *Oncogene, 23*, 4413–4421.
110. Payne, L. N. (1992). Developments in avian leukosis research. *Leukemia, 6*(Suppl 3), 150S–152S.
111. Payne, G. S., Bishop, J. M., & Varmus, H. E. (1982). Multiple arrangements of viral DNA and an activated host oncogene in bursal lymphomas. *Nature, 295*, 209–214.
112. Payne, G. S., Courtneidge, S. A., Crittenden, L. B., Fadly, A. M., Bishop, J. M., & Varmus, H. E. (1981). Analysis of avian leukosis virus DNA and RNA in bursal tumours: Viral gene expression is not required for maintenance of the tumor state. *Cell, 23*, 311–322.
113. Payne, L. N., & Fadly, A. M. (1997). Leukosis/sarcoma group. In: B. W. Calnek, H. J. Barnes, C. W. Beard, L. R. McDougald, & Y. M. Saif, (Ed.), *Diseases of Poultry* (10th ed., pp. 414–466). Ames, IA: Iowa State University Press.
114. Pecenka, V., Dvorak, M., Karafiat, V., Sloncova, E., Hlozanek, I., Travnicek, M., et al. (1988a). Avian nephroblastomas induced by a retrovirus (MAV-2) lacking oncogene. II. Search for common sites of proviral integration in tumour DNA. *Folia Biologica (Praha), 34*, 147–169.
115. Pecenka, V., Dvorak, M., Karafiat, V., & Travnicek, M. (1988b). Avian nephroblastomas induced by a retrovirus (MAV-2) lacking oncogene. III. Presence of defective MAV-2 proviruses in tumour DNA. *Folia Biologica (Praha), 34*, 215–232.
116. Perkins, K. J., Lusic, M., Mitar, I., Giacca, M., & Proudfoot, N. J. (2008). Transcription-dependent gene looping of the HIV-1 provirus is dictated by recognition of pre-mRNA processing signals. *Molecular Cell, 29*, 56–68.
117. Pizer, E. S., Baba, T. W., & Humphries, E. H. (1992). Activation of the c-myb locus is insufficient for the rapid induction of disseminated avian B-cell lymphoma. *Journal of Virology, 66*, 512–523.

118. Pizer, E., & Humphries, E. H. (1989). RAV-1 insertional mutagenesis: Disruption of the c-myb locus and development of avian B-cell lymphomas. *Journal of Virology*, *63*, 1630–1640.
119. Polony, T. S., Bowers, S. J., Neiman, P. E., & Beemon, K. L. (2003). Silent point mutation in an avian retrovirus RNA processing element promotes c-myb-associated short-latency lymphomas. *Journal of Virology*, *77*, 9378–9387.
120. Pulciani, S., Santos, E., Long, L. K., Sorrentino, V., & Barbacid, M. (1985). ras gene Amplification and malignant transformation. *Molecular and Cellular Biology*, *5*, 2836–2841.
121. Purchase, H. G., Okazaki, W., Vogt, P. K., Hanafusa, H., Burmester, B. R., & Crittenden, L. B. (1977). Oncogenicity of avian leukosis viruses of different subgroups and of mutants of sarcoma viruses. *Infection and Immunity*, *15*, 423–428.
122. Quintrell, N., Hughes, S. H., Varmus, H. E., & Bishop, J. M. (1980). Structure of viral DNA and RNA in mammalian cells infected with avian sarcoma virus. *Journal of Molecular Biology*, *143*, 363–393.
123. Raines, M. A., Lewis, W. G., Crittenden, L. B., & Kung, H. J. (1985). c-erbB activation in avian leukosis virus-induced erythroblastosis: Clustered integration sites and the arrangement of provirus in the c-erbB alleles. *Proceedings of the National Academy of Sciences of the USA*, *82*, 2287–2291.
124. Raines, M. A., Maihle, N. J., Moscovici, C., Crittenden, L., & Kung, H. J. (1988). Mechanism of c-erbB transduction: Newly released transducing viruses retain poly(A) tracts of erbB transcripts and encode C-terminally intact erbB proteins. *Journal of Virology*, *62*, 2437–2443.
125. Rangarajan, A., & Weinberg, R. A. (2003). Opinion: Comparative biology of mouse versus human cells: Modelling human cancer in mice. *Nature Reviews Cancer*, *3*, 952–959.
126. Robinson, H. L., & Gagnon, G. C. (1986). Patterns of proviral insertion and deletion in avian leukosis virus-induced lymphomas. *Journal of Virology*, *57*, 28–36.
127. Robinson, H. L., Miles, B. D., Catalano, D. E., Briles, W. E., & Crittenden, L. B. (1985). Susceptibility to erbB-induced erythroblastosis is a dominant trait of 151 chickens. *Journal of Virology*, *55*, 617–622.
128. Roebroek, A. J., Schalken, J. A., Onnekink, C., Bloemers, H. P., & Van de Ven, W. J. (1987). Structure of the feline c-fes/fps proto-oncogene: Genesis of a retroviral oncogene. *Journal of Virology*, *61*, 2009–2016.
129. Ryden, T. A., de Mars, M., & Beemon, K. (1993). Mutation of the C/EBP binding sites in the Rous sarcoma virus long terminal repeat and gag enhancers. *Journal of Virology*, *67*, 2862–2870.
130. Sage, J., Miller, A. L., Perez-Mancera, P. A., Wysocki, J. M., & Jacks, T. (2003). Acute mutation of retinoblastoma gene function is sufficient for cell cycle re-entry. *Nature*, *424*, 223–228.
131. Sauvageau, M., Miller, M., Lemieux, S., Lessard, J., Hebert, J., & Sauvageau, G. (2008). Quantitative expression profiling guided by common retroviral insertion sites reveals novel and cell type specific cancer genes in leukemia. *Blood*, *111*, 790–799.
132. Schubach, W. H., & Horvath, G. (1988). Alternate structures and stabilities of c-myc RNA in a bursal lymphoma cell line. *Nucleic Acids Research*, *16*, 11171–11186.
133. Selten, G., Cuypers, H. T., & Berns, A. (1985). Proviral activation of the putative oncogene Pim-1 in MuLV induced T-cell lymphomas. *EMBO Journal*, *4*, 1793–1798.
134. Shaughnessy, J. D., Jr., Owens, J. D., Jr., Wiener, F., Hilbert, D. M., Huppi, K., Potter, M., et al. (1993). Retroviral enhancer insertion 5′ of c-myc in two translocation-negative mouse plasmacytomas upregulates c-myc expression to different extents. *Oncogene*, *8*, 3111–3121.
135. Shtivelman, E., Zakut, R., & Canaani, E. (1984). Frequent generation of nonrescuable reorganized Moloney murine sarcoma viral genomes. *Proceedings of the National Academy of Sciences of the USA*, *81*, 294–298.

136. Simon, M. C., Neckameyer, W. S., Hayward, W. S., & Smith, R. E. (1987). Genetic determinants of neoplastic diseases induced by a subgroup F avian leukosis virus. *Journal of Virology*, *61*, 1203–1212.
137. Simon, M. C., Smith, R. E., & Hayward, W. S. (1984). Mechanisms of oncogenesis by subgroup F avian leukosis viruses. *Journal of Virology*, *52*, 1–8.
138. Smith, M. R., Smith, R. E., Dunkel, I., Hou, V., Beemon, K. L., & Hayward, W. S. (1997). Genetic determinant of rapid-onset B-cell lymphoma by avian leukosis virus. *Journal of Virology*, *71*, 6534–6540.
139. Sorensen, K. D., Quintanilla-Martinez, L., Kunder, S., Schmidt, J., & Pedersen, F. S. (2004). Mutation of all Runx (AML1/core) sites in the enhancer of T-lymphomagenic SL3-3 murine leukemia virus unmasks a significant potential for myeloid leukemia induction and favors enhancer evolution toward induction of other disease patterns. *Journal of Virology*, *78*, 13216–13231.
140. Soret, J., Dambrine, G., & Perbal, B. (1989). Induction of nephroblastoma by myeloblastosis-associated virus type 1: State of proviral DNAs in tumor cells. *Journal of Virology*, *63*, 1803–1807.
141. Starr, T. K., Allaei, R., Silverstein, K. A., Staggs, R. A., Sarver, A. L., Bergemann, T. L., et al. (2009). A transposon-based genetic screen in mice identifies genes altered in colorectal cancer. *Science*, *323*, 1747–1750.
142. Stewart, M., Mackay, N., Hanlon, L., Blyth, K., Scobie, L., Cameron, E., et al. (2007). Insertional mutagenesis reveals progression genes and checkpoints in MYC/Runx2 lymphomas. *Cancer Research*, *67*, 5126–5133.
143. Stoye, J. P., Moroni, C., & Coffin, J. M. (1991). Virological events leading to spontaneous AKR thymomas. *Journal of Virology*, *65*, 1273–1285.
144. Stratton, M. R., Campbell, P. J., & Futreal, P. A. (2009). The cancer genome. *Nature*, *458*, 719–724.
145. Suzuki, T., Shen, H., Akagi, K., Morse, H. C., Malley, J. D., Naiman, D. Q., et al. (2002). New genes involved in cancer identified by retroviral tagging. *Nature Genetics*, *32*, 166–174.
146. Swanstrom, R., Parker, R. C., Varmus, H. E., & Bishop, J. M. (1983). Transduction of a cellular oncogene: The genesis of Rous sarcoma virus. *Proceedings of the National Academy of Sciences of the USA*, *80*, 2519–2523.
147. Swift, R. A., Boerkoel, C., Ridgway, A., Fujita, D. J., Dodgson, J. B., & Kung, H. J. (1987). B-lymphoma induction by reticuloendotheliosis virus: Characterization of a mutated chicken syncytial virus provirus involved in c-myc activation. *Journal of Virology*, *61*, 2084–2090.
148. Theodorou, V., Kimm, M. A., Boer, M., Wessels, L., Theelen, W., Jonkers, J., et al. (2007). MMTV insertional mutagenesis identifies genes, gene families and pathways involved in mammary cancer. *Nature Genetics*, *39*, 759–769.
149. Tsichlis, P. N., & Coffin, J. M. (1980). Recombinants between endogenous and exogenous avian tumor viruses: Role of the C region and other portions of the genome in the control of replication and transformation. *Journal of Virology*, *33*, 238–249.
150. Uren, A. G., Kool, J., Berns, A., & van Lohuizen, M. (2005). Retroviral insertional mutagenesis: Past, present and future. *Oncogene*, *24*, 7656–7672.
151. Van Dyck, F., Declercq, J., Braem, C. V., & Van de Ven, W. J. (2007). PLAG1, the prototype of the PLAG gene family: Versatility in tumour development (review). *International Journal of Oncology*, *30*, 765–774.
152. van Lohuizen, M., Breuer, M., & Berns, A. (1989). N-myc is frequently activated by proviral insertion in MuLV-induced T cell lymphomas. *EMBO Journal*, *8*, 133–136.
153. Venugopal, K. (1999). Avian leukosis virus subgroup J: A rapidly evolving group of oncogenic retroviruses. *Research in Veterinary Science*, *67*, 113–119.
154. Virchow, R. (1863). Aetiologie der neoplastischen Geschwǀste/Pathogenie der neoplastischen Geschwǀste. In: R. Virchow, (Ed.), *Die krankhaften Geschwǀste* (pp. 57–101). Berlin: Verlag von August von Hirschwald.

155. Vogt, P. K. (1971). Spontaneous segregation of nontransforming viruses from cloned sarcoma viruses. *Virology*, *46*, 939–946.
156. Wagh, P. K., Peace, B. E., & Waltz, S. E. (2008). Met-related receptor tyrosine kinase Ron in tumor growth and metastasis. *Advance in Cancer Research*, *100*, 1–33.
157. Wang, W., Warren, M., & Bradley, A. (2007). Induced mitotic recombination of p53 in vivo. *Proceedings of the National Academy of Sciences of the USA*, *104*, 4501–4505.
158. Watts, S. L., & Smith, R. E. (1980). Pathology of chickens infected with avian nephoblastoma virus MAV-2(N. *Infection and Immunity*, *27*, 501–512.
159. West, A. G., & Fraser, P. (2005). Remote control of gene transcription. *Human Molecular Genetics*, *14*(Spec No 1), R101–R111.
160. Westaway, D., Papkoff, J., Moscovici, C., & Varmus, H. E. (1986). Identification of a provirally activated c-Ha-ras oncogene in an avian nephroblastoma via a novel procedure: cDNA cloning of a chimaeric viral-host transcript. *EMBO Journal*, *5*, 301–309.
161. Wong, S., & Witte, O. N. (2004). The BCR-ABL story: Bench to bedside and back. *Annual Review of Immunology*, *22*, 247–306.
162. Wood, L. D., Parsons, D. W., Jones, S., Lin, J., Sjoblom, T., Leary, R. J., et al. (2007). The genomic landscapes of human breast and colorectal cancers. *Science*, *318*, 1108–1113.
163. Wu, X., Luke, B. T., & Burgess, S. M. (2006). Redefining the common insertion site. *Virology*, *344*, 292–295.
164. Yamamoto, T., de Crombrugghe, B., & Pastan, I. (1980). Identification of a functional promoter in the long terminal repeat of Rous sarcoma virus. *Cell*, *22*, 787–797.
165. Yang, F., Xian, R. R., Li, Y., Polony, T. S., & Beemon, K. L. (2007). Telomerase reverse transcriptase expression elevated by avian leukosis virus integration in B cell lymphomas. *Proceedings of the National Academy of Sciences of the USA*, *104*, 18952–18957.
166. Yant, S. R., Wu, X., Huang, Y., Garrison, B., Burgess, S. M., & Kay, M. A. (2005). High-resolution genome-wide mapping of transposon integration in mammals. *Molecular and Cellular Biology*, *25*, 2085–2094.
167. Yim, E. K., Peng, G., Dai, H., Hu, R., Li, K., Lu, Y., et al. (2009). Rak functions as a tumor suppressor by regulating PTEN protein stability and function. *Cancer Cell*, *15*, 304–314.
168. Zhang, J., & Ma, Y. (2001). Evidence for retroviral intramolecular recombinations. *Journal of Virology*, *75*, 6348–6358.
169. Zhang, Y., & Rowley, J. D. (2006). Chromatin structural elements and chromosomal translocations in leukemia. *DNA Repair (Amst)*, *5*, 1282–1297.
170. Zhou, J., & Levine, M. (1999). A novel cis-regulatory element, the PTS, mediates an anti-insulator activity in the Drosophila embryo. *Cell*, *99*, 567–575.
171. Zhu, Y., Ghosh, P., Charnay, P., Burns, D. K., & Parada, L. F. (2002). Neurofibromas in NF1: Schwann cell origin and role of tumor environment. *Science*, *296*, 920–922.

Chapter 5
Sleeping Beauty Models of Cancer

Jesse D. Riordan, Laura M. Rogers, Katherine E. Berquam-Vrieze, and Adam J. Dupuy

5.1 Introduction

The many advantages of retroviral insertional mutagenesis have been discussed in the previous chapters, and this strategy has played a significant role in furthering our current understanding of the genetic basis of cancer. However, retroviral insertional mutagenesis has two main limitations that have prevented this approach from being applied to many forms of cancer. First, naturally-occurring slow transforming retroviruses have a restricted cellular tropism within the infected host, thus limiting the types of cells that can be mutagenized by these viruses [29]. Second, retroviruses require the host cell to undergo mitosis in order to gain access to the nuclear genome and integrate as a provirus. The combined effects of these drawbacks have limited the application of retroviral insertional mutagenesis to the study of hematopoietic malignancies and mammary cancer in mice.

Recent work has begun to exploit another potential insertional mutagen in mice-transposable elements (i.e. transposons). Transposons have proven to be valuable in the genetic analysis of invertebrate organisms and have been used in forward genetic screens for nearly 30 years [2, 19]. Unfortunately, similar approaches were not available for the study of mammalian genetics due to the lack of active transposons in these species. However, this changed with the development of Sleeping Beauty—a reverse engineered Tc1/mariner transposon that was shown to function in vertebrate cells [20]. This chapter describes the recent application of the Sleeping Beauty transposon for cancer gene discovery in the mouse.

J.D. Riordan (✉)
Department of Anatomy and Cell Biology, University of Iowa, Iowa City, IA 52242, USA
e-mail: jesse-riordan@uiowa.edu

A.J. Dupuy, D.A. Largaespada (eds.), *Insertional Mutagenesis Strategies in Cancer Genetics*, DOI 10.1007/978-1-4419-7656-7_5,

5.2 Resurrecting a DNA Fossil: the Sleeping Beauty Transposon System

The Sleeping Beauty (SB) transposon was resurrected from an ancient Tc1/mariner superfamily of "cut and paste" DNA transposons (i.e. class II transposable elements) cloned from salmonid fish [20]. The individual transposons were inactive due to the acquisition of mutations over evolutionary time. However, Ivics and colleagues speculated that the sequence of the active ancestral element could be deduced by deriving a consensus sequence from the alignment of many independent Tc1-like transposable elements (TCEs) from several different species of fish, each having acquired different inactivating mutations. By reverting these mutations, they were able to awaken Sleeping Beauty from its slumber and restore its ability to function as a transposable element. Subsequent work has shown the SB transposon to be active in numerous cell types from a wide variety of vertebrate organisms including zebrafish, medaka, Xenopus, rat, mouse and human [9, 16, 21, 27, 33, 44].

5.3 Transposase/Transposon Structure

Like all cut-and-paste transposons, the SB system consists of two functional parts: the transposase enzyme and the transposon vector. The SB transposase has several conserved domains that are critical for its function. At the N-terminus of the transposase is a bipartite DNA-binding domain [22]. This bipartite domain consists of a PAI domain and an RED domain, which are bridged by a small GRRR domain that is similar to an AT-hook. The PAI and RED domains confer the specificity by which the transposase is able to recognize its target [8]. The C-terminal catalytic domain, which mediates the cleavage and joining reactions of the target DNA, is a DDE motif—so named for the two invariable aspartic acid and single glutamic acid residues which define this domain. The DDE catalytic domain is a common feature in transposase, recombinase and viral integrase enzymes.

As previously mentioned, the work of Ivics and colleagues generated the first active form of the SB transposase, called SB10 [20]. While the SB10 transposase was shown to have activity in a number of cell lines [20, 21, 33] as well as in mice [11, 13, 17], the transposition efficiency often limited the applications of the SB system. Thus a number of independent groups made additional modifications to the SB10 transposase in an effort to increase its efficiency. These efforts produced three improved versions of the SB transposase: SB11 [15], HSB17 [1] and SB100X [35] which have an activity 10–100 fold greater than that of SB10.

The second part of the SB system, the transposon vector, has a structure similar to that of other class II transposable elements. The boundaries at the 5′ and 3′ ends of an SB transposon are defined by the presence of an inverted repeat (IR), also referred to as an inverted terminal repeat (ITR). A specific right (IRR) and left (IRL) inverted repeat are required in order for a SB transposon to be functional [7]. Each

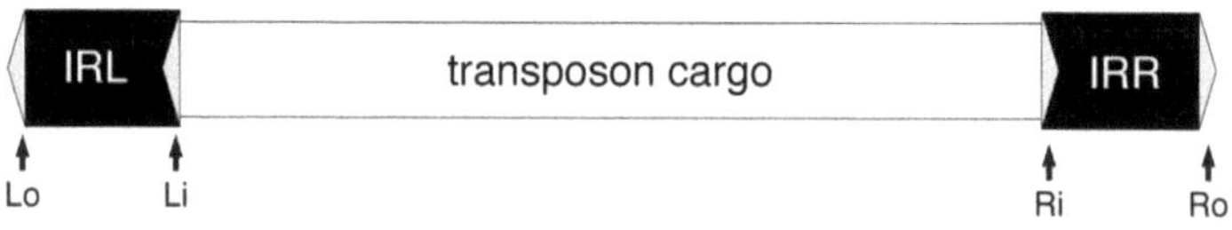

Lo - CAGTTGAAG**TCGGAAGTTTACATACAC**TTAAG
Li - TCCAGTGGG**TCAGAAGTTTACATACAC**TAAG

Ri - CCCAGTGGG**TCAGAAGTTTACATACAC**TCAAT
Ro - CAGTTGAAG**TCGGAAGTTTACATACAC**CTTAG

Fig. 5.1 Structure of Sleeping Beauty transposons. Each transposon consists of a DNA fragment (i.e. transposon cargo) flanked by a left (IRL) and right (IRR) inverted repeat. Each repeat is a unique sequence that is required for efficient transposition. Within each inverted repeat is an inner and outer direct repeat. The direct repeats define the 5′ and 3′ ends of each inverted repeat, as indicated. While the sequence of each direct repeat is unique (*shown below*), each contains a conserved sequence which represents the core binding sequence (*shown in bold*) of SB transposase

inverted repeat contains two direct repeats which serve as the binding sites for the SB transposase (Fig. 5.1). The direct repeats are unique, and all four are required for efficient transposition [7].

The initial transposons used to demonstrate SB function by Ivics et al. were derived from a cloned Tc1/mariner element isolated from *Tanichthys albonubes* (referred to as "T") [20]. This element differed from the consensus sequence by only 3.8%; while functional, it was not initially known how significantly these slight changes in the IR regions might reduce the transposition rate. Cui and colleagues later compared the sequence of the original "T" vector to a consensus sequence generated from the alignment of multiple Tc1/mariner IRs. This comparison identified a number of candidate mutations in the "T" vector. A new transposon vector, called "T2", was then generated by reverting these mutations back to the Tc1/mariner consensus sequence. The cumulative effect of these changes led to a four-fold increase in the transposition rate when using a T2 vector [7]. Thus, modification of both the transposase enzyme and transposon vector can have an effect on the overall transposition efficiency of the SB system.

5.4 Molecular Characteristics of SB Transposition

5.4.1 Transposition Mechanism

In order to utilize the SB system as a genetic tool, it is important to consider the molecular characteristics of the system so that potential sources of bias can be taken into account. An ideal transposable element to be used as an insertional mutagen is one that inserts efficiently, yet randomly into the host cell genome without introducing non-specific mutations (i.e. not transposon-tagged). This type of transposon would be a flexible genetic tool that could be used to perform an unbiased, forward genetic screen.

Following the work of Ivics et al., a number of groups began to study the molecular mechanisms that govern the activity of Sleeping Beauty. Luo et al. performed an experiment to determine the molecular characteristics of SB transposition by creating a mouse embryonic cell line in which a single SB transposon containing a neomycin resistance marker was used to disrupt the expression of a puromycin drug resistance gene [33]. This experimental design had two key features that provided several insights into the mechanism of SB transposition. First, the use of a dual drug-resistance screen allowed both the excision and insertion events to be recovered for additional study (excision→puromycin resistance, integration→puromycin/neomycin resistance). Second, this experiment was the first to induce transposition of an element that was already resident on a host cell chromosome. This is in contrast to Ivics et al. in which the transposons were introduced as plasmids [20].

The results of this experiment made two major contributions to our understanding of the SB system. First, analysis of the donor site showed that the SB transposase leaves behind a 3 basepair remnant in between the TA dinucleotides that flank the transposon prior to excision. This remnant is called a "footprint" and is common to the Tc1/mariner family [38]. More importantly, the presence of a footprint at the donor site indicates that the transposition mechanism used by the SB transposase

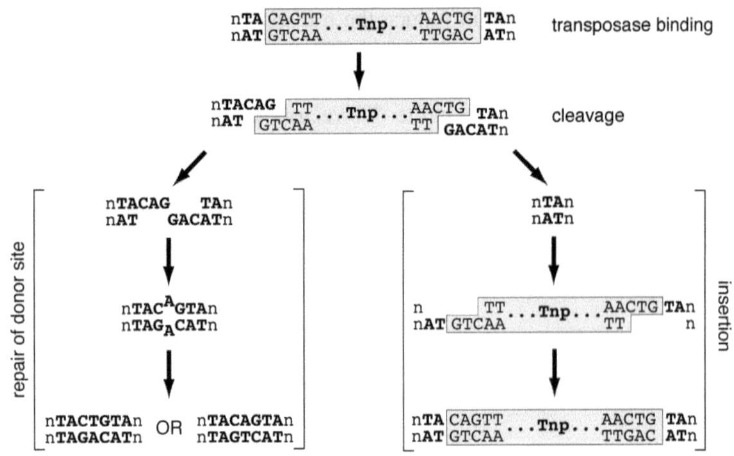

Fig. 5.2 Proposed mechanism of Sleeping Beauty transposition. Once a molecule of SB transposase binds to each of the four direct repeats (see Fig. 5.1), the transposase enzymes mediate DNA cleavage at the end of each inverted repeat. This cleavage event generates a double-stranded break at the donor site which is repaired by the host cell. The SB transposase cleavage event leaves behind single-stranded overhangs that must be resolved in one of two ways, as indicated. SB transposase mediates insertion at a new TA dinucleotide site elsewhere in the host cell genome. The incorporation of the transposon at this site leads to duplication of the TA target site which then flanks the newly inserted transposon

is likely similar to that of other members of the Tc1/mariner family (Fig. 5.2). Subsequent work by two independent groups showed that the SB transposition mechanism is aided by a number of host cell proteins, including members of the non-homologous end-joining pathway [23, 42] as well as the DNA-bending protein, Hmgb1 [45]. A second observation made by Luo et al. showed that SB exhibits a tendency to integrate into target sites physically linked to the donor site. This phenomenon is often referred to as "local hopping". This topic will be discussed in the following section, since local hopping is the main source of integration bias in the Sleeping Beauty transposon system.

5.4.2 Integration Site Bias

All integrating vectors, be they viral or transposon, have some degree of insertion site bias. Recent work has shown that many viruses (e.g. MuLV, HIV) preferentially integrate within or near actively transcribed genes [36]. Similarly, studies of transposable elements in mammalian cells have shown that not all target sequences are equally likely to be used by some transposons (e.g. piggyBac); and thus these transposons show a comparable bias for insertion near genes [31]. While the exact molecular cause of this type of integration bias has not been elucidated, the implication is that the integration mechanism of these vectors somehow recognizes the chromatin structure associated with active transcription. A more practical consequence of integration bias is that the mutagenic potential of these vectors is limited to actively transcribed regions in the host cell.

By contrast, an increasing number of publications have shown that the Sleeping Beauty transposon does not display a strong integration bias. To date, analysis of over 1800 SB transposon insertions identified in mouse hepatocytes [43], mouse ES cells [31] and mouse embryos [10] have shown that SB does not show a significant bias for insertion near genes. Instead, SB transposon insertions show a near random distribution with the percentage of insertion events within genes roughly equivalent to the percentage of the mouse genome known to encode genes.

The reduced bias for insertion within genes makes SB unique among integrating vectors. However, this is not to say that SB does not show some local preferences for specific insertion sites. For instance, several groups have analyzed large numbers of unselected SB insertions from a variety of sources to identity any integration site preferences that the SB transposase displays. The results show that SB does prefer AT-rich sequences flanking the TA target site [32, 41]. Another study examined DNA structure to look for any trends in the integration sites utilized by SB. This analysis revealed that SB has a preference for target sites in "bendable" regions [14]. It should be noted that all of these studies identified integration preferences for SB; while these factors may determine which target site (i.e. TA dinucleotide) in a local region will be used by the transposase, the cumulative effects of these preferences do not appear to cause significant integration site bias at the genomic level.

5.4.3 Local Hopping

As mentioned previously, Luo et al. was the first to demonstrate that SB exhibits local hopping [33]. Local hopping is a phenomenon common to cut-and-paste transposons including P elements [40] and Tol2 [25]. Additional experiments using the SB system in transgenic mice have confirmed the findings of Luo et al. in that all have reported some degree of local hopping. However, the frequency and the size of the genetic interval affected by local hopping varied significantly [4, 13, 18]. The source of this variation is not clear. It should be noted that these initial experiments were performed prior to the development of deep sequencing technology, and in some cases, prior to the sequencing of the mouse genome. As a result, the limited amount of transposon integration data available may have introduced sampling bias in the analysis of these early experiments using SB in the mouse. Nevertheless, local hopping—regardless of the frequency—is the largest source of integration site bias identified with the Sleeping Beauty transposon system.

5.4.4 Integration Site Mapping

Perhaps the greatest advantage of insertional mutagenesis is the ease with which the mutated genes can be identified. The earliest work with the SB system used a variety of PCR-based approaches (e.g. inverse PCR, splinkerette PCR) to amplify transposon junction fragments from the genomes of cells which had undergone SB mutagenesis. The PCR products were then cloned and sequenced. The junction sequences were sufficient to demonstrate transposase-mediated insertion into what was presumed to be the host cell genome. This presumption was based on two observations of the junction sequence: (1) the presence of a TA dinucleotide flanking the transposon and (2) junction sequence that was divergent from the initial donor site indicating the transposon had indeed hopped to another location.

Mapping transposon junction sequences was initially a laborious process that involved the use of radiation hybrid mapping [13] or fluorescence in situ hybridization [17] to obtain an approximate map location for each transposon insertion event. This process became easier with the completion of the mouse genome, and transposon junctions could be mapped to a precise chromosomal and nucleotide position. However, mapping transposon insertion sites still relied on sequences derived from cloned transposon junction fragments.

More recently, this process has been made even faster and more efficient with the development of next-generation sequencing technologies. These devices are capable of directly sequencing hundreds of thousands to millions of PCR products, thus eliminating the need to clone them. Hand curation of DNA sequences is no longer possible when processing such large data sets. Therefore, a number of groups have developed automated systems to analyze transposon insertion data generated by deep sequencing platforms [12, 28]. The combination of deep sequencing with these automated sequence analysis systems has provided an opportunity to study the process of SB transposition in much greater detail.

5.5 Sleeping Beauty in Cancer Research

Most of the initial publications that described the use of the SB system in mice focused on germline mutagenesis to isolate mutant strains with developmental phenotypes. Unfortunately, this approach was limited by the relatively low rate of SB mutagenesis compared to chemical mutagenesis using compounds such as ethylnitrosourea (ENU). This effort was also confounded by the high rate of local hopping which further reduced the efficiency of whole-genome mutagenesis using SB. Despite these limitations, several groups began to adapt the SB system to mutagenize somatic cells to induce tumors in mice. It was thought that the low rate of mutagenesis would be sufficient to induce tumors, given enough time. If so, the SB system could be used to perform forward genetic screens to identify cancer genes in mice.

Understanding which genes contribute to cancer has long been a major goal of cancer researchers. Next-generation DNA sequencing technologies have greatly reduced the cost and dramatically increased the pace at which mutation profiles from individual tumors can be generated [34]. Unfortunately, recent work suggests that human tumors have a mutator phenotype with a mutation rate ~200 fold greater than normal tissue [3]. As a consequence, the precise identification of the mutations which contribute to the cancer phenotype (i.e. driver mutations) among the more abundant background mutations (i.e. passenger mutations) is a major challenge to the field of cancer genetics. However, Sleeping Beauty transposon mutagenesis could prove to be a useful tool for carrying out forward genetic screens in mice, allowing researchers to more rapidly identify somatic mutations that promote cancer in vivo. By combining data from these SB screens with sequence data from human cancer genomes, perhaps driver mutations, and thus more appropriate therapeutic targets, could be identified.

5.5.1 Modifications of Transposon Design

Initial efforts using the SB system as a germline insertional mutagen in mice made use of gene trap vectors that would express a reporter gene (e.g. EGFP, β-gal) upon transposon insertion within a transcription unit [4, 17]. This approach is useful for the identification and isolation of individual transposon-induced mutations, given the inefficiency of the SB system in this context. However, since cancer exhibits a strong positive selection in vivo, reporter genes have little utility. Moreover, several groups have shown that the efficiency of SB-induced transposition is greatly reduced as the transposon size increases [15, 24]. These factors influenced transposon design as the SB system was adapted to induce somatic cell insertional mutations to model cancer.

The T2/Onc vector was the first transposon to be specifically engineered for this purpose [6]. Its compact design includes the necessary elements to induce both gain and loss of function mutations in oncogenes and tumor suppressor genes, respectively (Fig. 5.3). The total size of T2/Onc is approximately 2 kb, which closely resembles the original size of the salmonid consensus sequence and seems to be the optimal size for transposition [15]. Splice acceptor sites and polyadenylation sites are included on both strands of the transposon, allowing this vector to function as a bi-directional gene trap. Together, these sequences are capable of disrupting tumor suppressor genes (Fig. 5.4a). Additionally, T2/Onc includes a promoter and splice donor which provide the ability to overexpress downstream oncogenes (Fig. 5.4b). It should be noted that several T2/Onc variants (T2/Onc2, T2/Onc3) have been developed [10, 12]. However, these transposon vectors share the same overall design and therefore share common mutational mechanisms.

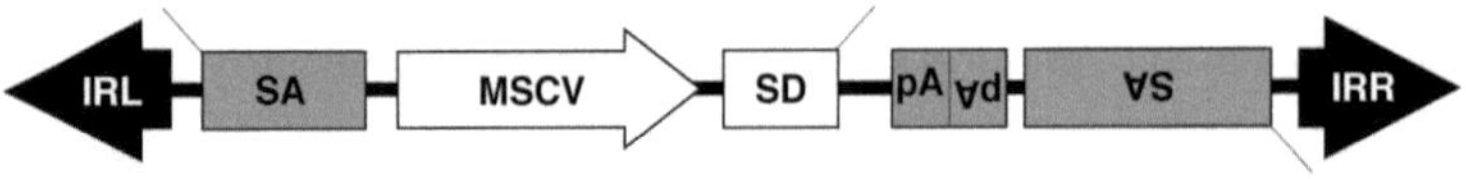

Fig. 5.3 Structure of the mutagenic T2/Onc transposon. The T2/Onc transposon is designed to induce mutations, but does not encode any reporter genes. Splice acceptors (SA) and polyadenylation sites (pA) function as gene trap cassettes to disrupt gene expression. The murine stem cell virus promoter (MSCV) and splice donor (SD) sequences act as a synthetic first exon to initiate transcription and splice into downstream genes

5.5.2 *Initial Tumor Models Induced by Transposition*

While most genetic diseases are caused by inherited mutations (i.e. present in every somatic cell), cancer is caused by the accumulation of mutations in a small population of somatic cells. The combination of spontaneous mutagenesis within normal cells and the constant selection for survival and growth contribute to tumor initiation and progression. This makes the Sleeping Beauty system ideal for mimicking human cancer in mice since transposition of mutagenic vectors (e.g. T2/Onc) produces unique combinations of mutations in each cell. Positive selection need only select the cell which harbors the right combination of mutations to emerge as a tumor. The major advantage of using this approach is that the mutations induced by SB mutagenesis will be tagged with transposons, thus facilitating their rapid identification.

Both parts of the SB system—transposase and transposon—must be introduced into mice for transposition to occur. The first publications describing the use of the SB system as a cancer gene discovery tool described two different approaches [6, 10]. Collier et al. made use of an existing transgenic mouse strain (CAGGS-SB10) to ubiquitously express the SB10 transposase [11]. This allele was combined

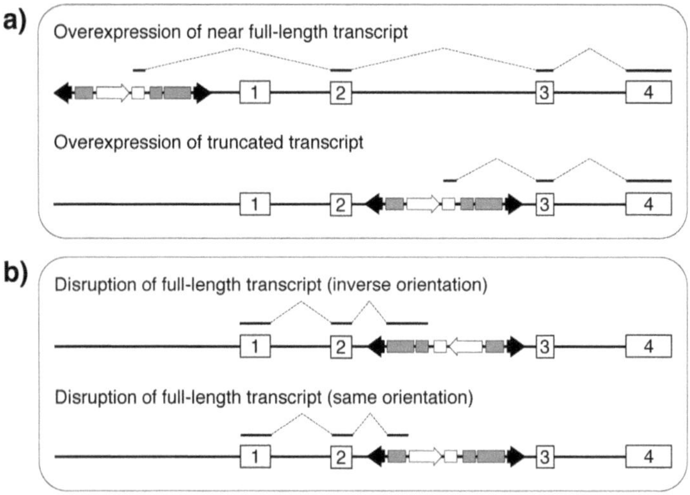

Fig. 5.4 Mechanisms of mutation induced by T2/Onc transposons in SB-induced tumors. **a** Overexpression of oncogenes in SB-induced tumors is accomplished through two main mechanisms in which the T2/Onc transposon overexpresses a near full-length (*above*) or truncated (*below*) transcript. These mechanisms employ the MSCV promoter and splice donor within the transposon (see Fig. 5.3). **b** Disruption of tumor suppressor genes in SB-induced tumors is typically achieved through the action of the gene trap elements on the plus strand (*above*) or minus strand (*below*), depending on the orientation of the transposon relative to the mutated gene

with a transgene consisting of multiple copies of the mutagenic T2/Onc transposon. While double transgenic mice (i.e. CAGGS-SB10$^+$;T2/Onc$^+$) did not develop spontaneous tumors in a wild type background, this combination of SB alleles dramatically accelerated tumor formation in p19-null mice. Molecular characterization of sarcomas from these mice indicated that transposon-induced mutations in the *Braf* locus were likely responsible for the decreased tumor latency [6]. This was an important demonstration that the SB system could be used to study the genetics of solid tumors.

A second publication described two novel SB alleles [10]. The first was a knock-in mouse strain (RosaSBase) which expresses the optimized SB11 transposase from the ubiquitously-transcribed *ROSA26* locus. Dupuy et al. also generated several transgenic mouse strains harboring the T2/Onc2 transposon. This mutagenic vector is virtually identical to T2/Onc, except its size has been further reduced to increase transposition efficiency. Another important distinction of the T2/Onc2 transgenic mice is that each cell contains over a hundred copies of the T2/Onc2 transposon. This increase in the transposon copy number per cell likely increases the number of transposon-induced mutations that can be achieved.

Unlike the results reported by Collier et al., offspring inheriting both RosaSBase and T2/Onc2 commonly developed T-cell lymphomas and other hematopoietic tumors [10]. These tumors developed in otherwise wild type mice and did not require a sensitizing mutation, such as the p19-null background used by Collier et al. However, the combination of RosaSBase and T2/Onc2 had a major limitation in that the majority of double transgenic mice died at midgestation, likely due to active transposition. Additionally, mice that survived until birth developed predominantly T-cell lymphoma rather than a diverse array of tumor types. Nevertheless, this work showed that the efficiency of the SB system was sufficient to induce tumors in wild type mice.

The origin of the apparent discrepancy in transposition efficiencies of the SB models described by Collier et al. and Dupuy et al. was not immediately clear. The improved efficiency described in RosaSBase;T2/Onc2 double transgenic mice could be due to enhanced transposase expression from the *ROSA26* locus, increased copy number of the mutagenic transposon vector or a combination of both. Recent work by Collier et al. has shown that both transposon and transposase alleles make a contribution to the efficiency of the system [5]. These experiments combined the RosaSBase allele with the lower copy T2/Onc transgenic strains. The resulting double transgenic mice developed tumors without the requirement of a sensitizing mutation (e.g. $p19^{-/-}$), suggesting that the RosaSBase allele drives a higher rate of transposition, compared to the CAGGS-SB10 allele. However, RosaSBase;T2/Onc mice did not display an embryonic lethal phenotype and developed tumors with an increased latency compared to RosaSBase;T2/Onc2 mice [5]. This result indicates that the increased transposon copy number also contributes to transposition rate. Thus the nature of both the transposase and transposon components affect the tumor phenotype in SB models of cancer.

5.5.3 Identification of Driver Mutations in SB-Induced Tumors

As noted earlier, a number of PCR-based approaches have recently been described to identify transposon integration sites in SB-induced tumors [5, 12, 26, 39]. Direct sequencing of PCR products using next generation sequencing platforms has greatly improved the sensitivity of detection for transposon-induced mutations in these tumors. However, the increased scale of sequencing has also uncovered significant genetic complexity in SB-induced tumors. While this complexity is a necessary component of a good model of human cancer, it also produces a similar problem that has been encountered with the sequencing of human cancer genomes—how are rare driver mutations identified among the large number of passenger mutations?

This problem can easily be addressed when using the SB system. As previously noted, the SB transposase exhibits very weak integration site bias. Therefore, unselected SB insertion events can be modeled in silico by randomly selecting TA dinucleotide sites throughout the genome. Currently, this is achieved using a Monte Carlo simulation to identify common integration sites (CISs)—regions of the genome that suffer transposon insertion in SB-induced tumors at a rate more frequently predicted by chance [5, 12, 26, 39]. Significance is a function of the number of tumors having the observed number of integration events within a specified genomic region (see Chapter 7 for more detail). It is important to note that such methods cannot easily be employed in the identification of driver mutations in human cancer since spontaneous mutations do not occur randomly and are influenced by a number of factors.

One caveat in the identification of driver mutations in SB-induced tumors involves local hopping—the main source of integration bias in the SB system. There is not a well-established definition of the local hopping interval for the SB system currently in use. In fact, it appears that the rate of transposon insertion is elevated along the entire length of the chromosome that harbors the transposon transgene (our unpublished data). Consequently, the genes which map to this chromosome are subject to an increased likelihood of transposon-induced mutation compared to the rest of the genome. Currently, the only method to prevent the identification of false-positive CISs due to local hopping in SB models of cancer is to discard all integration events which map to the local chromosome. While this approach leads to the loss of data from an entire chromosome in each transposon strain, multiple mouse strains harboring transposon transgenes which map to different chromosomes can be used to compensate and to negate the loss of the local chromosome in each individual strain.

5.6 Modifications to the SB System

While the initial SB transposase (CAGGS-SB10, RosaSBase) and mutagenic transposon alleles (T2/Onc, T2/Onc2) showed that the SB system could be used to produce tumors in vivo, the earliest SB models were limited in the number of tumor types that could be produced. However, this limitation was unlikely due to the activity of the SB system, as previous work had observed transposition in a variety of tissue types [6, 10, 17]. Thus, it seemed likely that modifications to the existing transposase and transposon alleles could provide more precise control, allowing for tissue-specific transposon mutagenesis.

The structure of the mutagenic transposon has recently been investigated as one manner in which the SB system can be modified. Both the T2/Onc and T2/Onc2 transposons contain the murine stem cell virus (MSCV) promoter to drive overexpression of downstream oncogenes. This promoter is most active in cells of hematopoietic origin, which could explain why mutagenesis screens conducted with T2/Onc and T2/Onc2 transposons yielded primarily hematopoietic malignancies [5, 12]. In a modified version of the mutagenic transposon, called T2/Onc3, the MSCV promoter has been replaced with the cytomegalovirus early enhancer/chicken β-actin (CAG) promoter, which displays its highest activity in cells of epithelial origin [37]. The tumor profile obtained from a mutagenesis screen using T2/Onc3 transposons consisted mostly of carcinomas [12]. The ability to generate this type of profile is highly advantageous, as the majority of human malignancies are solid tumors. As evidenced by the different tumor profiles generated by these two independent screens, altering the promoter within mutagenic SB transposons can dramatically affect the nature of tumors resulting from integration events.

A potential application of this finding is the generation of transposons containing tissue-specific promoters whose activity is limited to a specific organ of interest. A mutagenesis screen utilizing this type of transposon would be predicted to generate tumors primarily in the organ of interest. This would allow rapid generation of a large number of independent tumors from that organ, which would facilitate in-depth analysis of the mutation profile of that specific tumor type. By performing independent mutagenesis screens with different transposons, novel oncogenes and tumor suppressors involved in tumorigenesis of a wide variety of tissue types could be identified.

A second approach to generate tissue-specific tumor models with the SB system is to limit transposition events, and therefore mutations, to the tissue of interest. This was recently achieved by generating a conditional transposase allele (RosaSBase-LsL) that is only expressed upon Cre recombinase-mediated removal of a Lox-stop-Lox (LsL) cassette [12]. As with constitutive SB models, the SB transposase gene is knocked into the *ROSA26* locus to allow its expression in any tissue of the mouse. Expression is prevented, however, by inclusion of an LsL cassette just upstream of the transposase coding region. This cassette, which consists of an EGFP cDNA followed by three polyadenylation signals, traps transcripts initiated from the *ROSA26* promoter and prevents expression of the downstream transposase gene. Thus, no transposition occurs in mice harboring both the conditional transposase allele and a transposon transgene due to a lack of transposase expression. Tissue-specific expression of Cre recombinase leads to excision of the LsL cassette, and thus transposase expression.

The RosaSBase-LsL allele has been used successfully to generate specific SB-induced carcinoma models. For example, [26] initiated transposon mutagenesis specifically in hepatocytes by expressing Cre from a transgene driven by the albumin promoter [26]. These mice develop preneoplastic liver nodules at ~160 days of age, some of which progress to hepatocellular carcinoma (HCC) as the mice age. Analysis of the transposon integration sites from these tumors identified a number of genes already associated with HCC (e.g. *Egfr, Met*) as well as a number of candidate genes not previously implicated in this disease (e.g. *Ube2h*). A second publication generated a model of colorectal cancer (CRC) by inducing transposition in the gastrointestinal tract using a Cre transgene expressed from the villin promoter [39]. Transposon mutagenesis produced multiple intestinal adenomas in approximately half of the mutagenized mice. As was the case with the SB-induced HCCs, analysis of transposon-induced mutations in these lesions identified a number of genes known to be involved in CRC, as well as a large number of candidate genes. These experiments are an important demonstration of the flexibility of the SB system to induce a broad spectrum of tumors in a controlled fashion. Moreover, the identification of genes known to be involved in human HCC and CRC in both of these screens implies that this is a valid approach to identify novel cancer genes.

5.7 Future Directions

5.7.1 Identifying Genes Involved in Metastasis

As described above, the SB insertional mutagenesis system provides a means to identify oncogenes and tumor suppressors that drive tumor formation and development in various tissues throughout the body. In addition to identifying genes involved in primary tumor formation, the system could be used to discover those that contribute to other aspects of tumor biology, such as metastasis. This approach would likely require a tissue-specific tumor model, as mice with ubiquitous transposition develop several tumors in multiple organs, which generally prevents them from surviving long enough for metastases to develop. Mice with tumorigenesis targeted to a single organ can survive longer, which may allow sufficient time for metastasis to occur, as was observed when transposition was limited to hepatocytes [26].

Once a model system has been established in which SB transposition drives formation of primary tumors that metastasize, the profile of transposon integration sites could be compared between primary tumors and their metastatic derivatives. Each metastatic tumor could be paired with the primary tumor from which it was derived based on similarities in overall integration profile. This type of analysis may lead to the identification of genes that are common sites of integration in metastases, but not in primary tumors. Depending on the nature of the mutation induced by transposon insertion, these genes would be implicated as either promoters or inhibitors of metastasis.

Another useful approach would be to compare integration profiles of primary tumors with identified metastases to those of primary tumors lacking metastases. Such an approach could lead to the discovery of gene disruptions in primary tumors that are predictive of metastatic potential. Additionally, genes may be identified that commonly contain transposon integrations in primary, but not metastatic tumors. This type of pattern would suggest that disruption of these genes may promote primary tumor formation, but may not be compatible with metastasis. Finally, analysis of integration profiles derived from tumors isolated from different metastatic sites (e.g. brain, lung, liver) may identify specific mutation events that promote metastasis to each distinct site. Such information could eventually prove useful in predicting patient outcome and treatment strategies.

5.7.2 Identifying Genes Involved in Treatment Response

A recent retroviral mutagenesis screen demonstrated the utility in using insertional mutagenesis to study treatment response in mice [30]. The SB system could be used in a similar manner to identify mutations that confer drug susceptibility or resistance. For this approach, a cohort of mice could be generated with SB-induced tumors. These mice could be treated with the drug of interest, then separated into two groups: those that respond to treatment and those that do not. Transposon integration profiles could be compared between responders and non-responders to identify mutations which are correlated with either resistance of susceptibility.

A modification to this approach could involve collection of tumor biopsies from animals before and after drug treatment. This would allow the investigator to follow genetic progression within an individual tumor in response to drug treatment. In this case, drug treatment would act as a selection event in which tumor cells that acquired mutations conferring drug resistance would preferentially replicate. Thus, transposon-induced mutations that appear to expand within the tumor mass following drug treatment would be candidate resistance mutations.

5.7.3 Identifying Cooperating Mutations

It is well established that cancer is the result of accumulated mutations which collectively work to transform a cell. As such, these mutations likely work in a synergistic fashion to modify signaling pathways or other subcellular processes. In most cases, however, it is not clear which specific mutations and/or pathways work in concert to contribute to cancer. Here again, SB models of cancer could be used to identify mutations that cooperate in generating the tumorigenic phenotype. In this scenario, transposon mutagenesis could be induced in mice carrying a predisposing mutation which serves as the initiating genetic event. Ongoing transposition would generate additional mutations promoting tumor progression. This approach could be utilized in the context of different predisposing mutations for a single tumor type. Distinct sets of cooperating genes may be identified for each initiating mutation by comparing transposon integration profiles derived from each model. Such information could lead to the identification of novel combinations of therapeutic targets for distinct subsets of tumors within a particular cancer type.

5.8 Challenges Remaining

As described in the previous sections, the current SB mutagenesis system has broad utility in the identification of genes associated with tumorigenesis and could potentially be used to study many other aspects of tumor biology. However, some challenges remain, which if properly addressed, would allow even broader applicability of the system.

Efforts to date have failed to efficiently generate tumors in certain organs. For example, a screen in which transposition was targeted to the prostate using the probasin-Cre strain failed to generate tumors. Similarly, targeting transposition to the mammary gland using the whey acidic protein (WAP)-Cre strain was not effective in generating mammary tumors (our unpublished data). At this point, it remains unclear whether this organ-specific resistance to SB-induced tumorigenesis results from a property of the SB system itself or from properties intrinsic to the target tissues. One possibility is that the promoters used to express Cre recombinase do not have the correct spatial and/or temporal expression pattern to drive mutagenesis in the appropriate initiating cell population for these forms of cancer. Consistent with this idea, mammary tumors have been induced using T2/Onc3 mutagenesis combined with ubiquitous transposition [12]. Another possibility deals with the activity of the transposon-encoded promoter. Changing the promoter within mutagenic transposons has been shown to dramatically affect the profile of tumors that are developed in a ubiquitous transposition model [12]. It is possible that the transposons used for tissue-specific prostate and mammary models do not efficiently initiate transcription in those tissues. Another factor that could influence tumor development in different tissues is the availability of host cell proteins used used by the SB transposase during transposition (e.g. Hmgb1) [45]. The expression level of these co-factors may be too low in some tissues to achieve the degree of mutagenesis required for transformation. Additional experiments are required to determine if the existing SB system will permit tissue-specific tumor models to be generated for these cancer types or if additional modifications will be necessary.

Another issue which has complicated the use of the Cre-inducible SB system is that many Cre transgenic mouse strains display some level of expression in multiple tissues. These Cre strains could produce a variety of tumor types when used to activate SB mutagenesis in mice. This would be problematic if Cre were expressed in the hematopoietic system, as the SB system is particularly efficient at inducing lymphomas. This issue could be partially addressed by combining the Cre-inducible transposition system with mutagenic transposons containing a promoter specific to the tissue of interest. While transposition may occur in multiple tissues, gain-of-function mutations, and thus tumor formation, would likely be limited to the tissue of interest.

All versions of the SB system currently in use to model cancer lack the ability to terminate transposon mutagenesis once it has been initiated. Achieving this goal would be advantageous for a number of reasons. First, it could facilitate the identification of genes associated with drug resistance or susceptibility. For example, transposon mobilization could be used to drive tumorigenesis and then shut off. Following this, drug treatment could be used to induce tumor regression. After significant regression, transposition could be reactivated, with the expectation that integrations causing disruption of genes conferring drug susceptibility would be selected for in the tumors. Additionally, the level of background integrations not contributing to tumor development or progression (i.e. passenger mutations) could be decreased by turning off transposition following initiation of tumorigenesis. Finally, the ability to turn transposition on and off would also allow a more in-depth analysis of distinct stages of tumor progression.

References

1. Baus, J., Liu, L., Heggestad, A. D., Sanz, S., & Fletcher, B. S. (2005). Hyperactive transposase mutants of the Sleeping Beauty transposon. *Molecular Therapy, 12*, 1148–1156.

2. Bazopoulou, D., & Tavernarakis, N. (2009). The NemaGENETAG initiative: large scale transposon insertion gene-tagging in Caenorhabditis elegans. *Genetica*, *137*, 39–46.
3. Bielas, J. H., Loeb, K. R., Rubin, B. P., True, L. D., & Loeb, L. A. (2006). Human cancers express a mutator phenotype. *Proceedings of the National Academy of Sciences of the USA*, *103*, 18238–18242.
4. Carlson, C. M., Dupuy, A. J., Fritz, S., Roberg-Perez, K. J., Fletcher, C. F., & Largaespada, D. A. (2003). Transposon mutagenesis of the mouse germline. *Genetics*, *165*, 243–256.
5. Collier, L. S., Adams, D. J., Hackett, C. S., Bendzick, L. E., Akagi, K., Davies, M. N., et al. (2009). Whole-body sleeping beauty mutagenesis can cause penetrant leukemia/lymphoma and rare high-grade glioma without associated embryonic lethality. *Cancer Research*, *69*, 8429–8437.
6. Collier, L. S., Carlson, C. M., Ravimohan, S., Dupuy, A. J., & Largaespada, D. A. (2005). Cancer gene discovery in solid tumours using transposon-based somatic mutagenesis in the mouse. *Nature*, *436*, 272–276.
7. Cui, Z., Geurts, A. M., Liu, G., Kaufman, C. D., & Hackett, P. B. (2002). Structure-function analysis of the inverted terminal repeats of the sleeping beauty transposon. *Journal of Molecular Biology*, *318*, 1221–1235.
8. Czerny, T., Schaffner, G., & Busslinger, M. (1993). DNA sequence recognition by Pax proteins: bipartite structure of the paired domain and its binding site. *Genes and Development*, *7*, 2048–2061.
9. Davidson, A. E., Balciunas, D., Mohn, D., Shaffer, J., Hermanson, S., Sivasubbu, S., et al. (2003). Efficient gene delivery and gene expression in zebrafish using the Sleeping Beauty transposon. *Developmental Biology*, *263*, 191–202.
10. Dupuy, A. J., Akagi, K., Largaespada, D. A., Copeland, N. G., & Jenkins, N. A. (2005). Mammalian mutagenesis using a highly mobile somatic Sleeping Beauty transposon system. *Nature*, *436*, 221–226.
11. Dupuy, A. J., Fritz, S., & Largaespada, D. A. (2001). Transposition and gene disruption in the male germline of the mouse. *Genesis*, *30*, 82–88.
12. Dupuy, A. J., Rogers, L. M., Kim, J., Nannapaneni, K., Starr, T. K., Liu, P., et al. (2009). A modified sleeping beauty transposon system that can be used to model a wide variety of human cancers in mice. *Cancer Research*, *69*, 8150–8156.
13. Fischer, S. E., Wienholds, E., & Plasterk, R. H. (2001). Regulated transposition of a fish transposon in the mouse germ line. *Proceedings of the National Academy of Sciences of the USA*, *98*, 6759–6764.
14. Geurts, A. M., Hackett, C. S., Bell, J. B., Bergemann, T. L., Collier, L. S., Carlson, C. M., et al. (2006). Structure-based prediction of insertion-site preferences of transposons into chromosomes. *Nucleic Acids Research*, *34*, 2803–2811.
15. Geurts, A. M., Yang, Y., Clark, K. J., Liu, G., Cui, Z., Dupuy, A. J., et al. (2003). Gene transfer into genomes of human cells by the sleeping beauty transposon system. *Molecular Therapy*, *8*, 108–117.
16. Grabher, C., Henrich, T., Sasado, T., Arenz, A., Wittbrodt, J., & Furutani-Seiki, M. (2003). Transposon-mediated enhancer trapping in medaka. *Gene*, *322*, 57–66.
17. Horie, K., Kuroiwa, A., Ikawa, M., Okabe, M., Kondoh, G., Matsuda, Y., et al. (2001). Efficient chromosomal transposition of a Tc1/mariner- like transposon Sleeping Beauty in mice. *Proceedings of the National Academy of Sciences of the USA*, *98*, 9191–9196.
18. Horie, K., Yusa, K., Yae, K., Odajima, J., Fischer, S. E., Keng, V. W., et al. (2003). Characterization of Sleeping Beauty transposition and its application to genetic screening in mice. *Molecular and Cellular Biology*, *23*, 9189–9207.
19. Hummel, T., & Klambt, C. (2008). P-element mutagenesis. *Methods in Molecular Biology*, *420*, 97–117.
20. Ivics, Z., Hackett, P. B., Plasterk, R. H., & Izsvak, Z. (1997). Molecular reconstruction of Sleeping Beauty, a Tc1-like transposon from fish, and its transposition in human cells. *Cell*, *91*, 501–510.

21. Izsvak, Z., Ivics, Z., & Plasterk, R. H. (2000). Sleeping beauty, a wide host-range transposon vector for genetic transformation in vertebrates. *Journal of Molecular Biology*, *302*, 93–102.
22. Izsvak, Z., Khare, D., Behlke, J., Heinemann, U., Plasterk, R. H., & Ivics, Z. (2002). Involvement of a bifunctional, paired-like DNA-binding domain and a transpositional enhancer in Sleeping Beauty transposition. *Journal of Biological Chemistry*, *277*, 34581–34588.
23. Izsvak, Z., Stuwe, E. E., Fiedler, D., Katzer, A., Jeggo, P. A., & Ivics, Z. (2004). Healing the wounds inflicted by sleeping beauty transposition by double-strand break repair in mammalian somatic cells. *Molecular Cell*, *13*, 279–290.
24. Karsi, A., Moav, B., Hackett, P., & Liu, Z. (2001). Effects of insert size on transposition efficiency of the sleeping beauty transposon in mouse cells. *Marine Biotechnology (New York)*, *3*, 241–245.
25. Keng, V. W., Ryan, B. J., Wangensteen, K. J., Balciunas, D., Schmedt, C., Ekker, S. C., et al. (2009a). Efficient transposition of Tol2 in the mouse germline. *Genetics*, *183*, 1565–1573.
26. Keng, V. W., Villanueva, A., Chiang, D. Y., Dupuy, A. J., Ryan, B. J., Matise, I., et al. (2009b). A conditional transposon-based insertional mutagenesis screen for genes associated with mouse hepatocellular carcinoma. *Nature Biotechnology*, *27*, 264–274.
27. Kitada, K., Ishishita, S., Tosaka, K., Takahashi, R., Ueda, M., Keng, V. W., et al. (2007). Transposon-tagged mutagenesis in the rat. *Nature Methods*, *4*, 131–133.
28. Kong, J., Zhu, F., Stalker, J., & Adams, D. J. (2008). iMapper: a web application for the automated analysis and mapping of insertional mutagenesis sequence data against Ensembl genomes. *Bioinformatics*, *24*, 2923–2925.
29. Kool, J., & Berns, A. (2009). High-throughput insertional mutagenesis screens in mice to identify oncogenic networks. *Nature Reviews Cancer*, *9*, 389–399.
30. Lauchle, J. O., Kim, D., Le, D. T., Akagi, K., Crone, M., Krisman, K., et al. (2009). Response and resistance to MEK inhibition in leukaemias initiated by hyperactive Ras. *Nature*, *461*, 411–414.
31. Liang, Q., Kong, J., Stalker, J., & Bradley, A. (2009). Chromosomal mobilization and reintegration of sleeping beauty and PiggyBac transposons. *Genesis*, *47*, 404–408.
32. Liu, G., Geurts, A. M., Yae, K., Srinivasan, A. R., Fahrenkrug, S. C., Largaespada, D. A., et al. (2005). Target site preferences of Sleeping Beauty transposons. *Journal of Molecular Biology*, *346*, 161–173.
33. Luo, G., Ivics, Z., Izsvak, Z., & Bradley, A. (1998). Chromosomal transposition of a Tc1/mariner like element in mouse embryonic stem cells. *Proceedings of the National Academy of Sciences of the USA*, *95*, 10769–10773.
34. Mardis, E. R., & Wilson, R. K. (2009). Cancer genome sequencing: a review. *Human Molecular Genetics*, *18*, R163–R168.
35. Mates, L., Chuah, M. K., Belay, E., Jerchow, B., Manoj, N., Acosta-Sanchez, A., et al. (2009). Molecular evolution of a novel hyperactive Sleeping Beauty transposase enables robust stable gene transfer in vertebrates. *Nature Genetics*, *41*, 753–761.
36. Mitchell, R. S., Beitzel, B. F., Schroder, A. R., Shinn, P., Chen, H., Berry, C. C., et al. (2004). Retroviral DNA integration: ASLV, HIV, and MLV show distinct target site preferences. *PLoS Biology*, *2*, E234.
37. Okabe, M., Ikawa, M., Kominami, K., Nakanishi, T., & Nishimune, Y. (1997). 'Green mice' as a source of ubiquitous green cells. *FEBS Letters*, *407*, 313–319.
38. Plasterk, R. H., Izsvak, Z., & Ivics, Z. (1999). Resident aliens: the Tc1/mariner superfamily of transposable elements. *Trends in Genetics*, *15*, 326–332.
39. Starr, T. K., Allaei, R., Silverstein, K. A., Staggs, R. A., Sarver, A. L., Bergemann, T. L., et al. (2009). A transposon-based genetic screen in mice identifies genes altered in colorectal cancer. *Science*, *323*, 1747–1750.
40. Tower, J., Karpen, G. H., Craig, N., & Spradling, A. C. (1993). Preferential transposition of Drosophila P elements to nearby chromosomal sites. *Genetics*, *133*, 347–359.

41. Vigdal, T. J., Kaufman, C. D., Izsvak, Z., Voytas, D. F., & Ivics, Z. (2002). Common physical properties of DNA affecting target site selection of sleeping beauty and other Tc1/mariner transposable elements. *Journal of Molecular Biology*, *323*, 441–452.
42. Yant, S. R., & Kay, M. A. (2003). Nonhomologous-end-joining factors regulate DNA repair fidelity during Sleeping Beauty element transposition in mammalian cells. *Molecular and Cellular Biology*, *23*, 8505–8518.
43. Yant, S. R., Wu, X., Huang, Y., Garrison, B., Burgess, S. M., & Kay, M. A. (2005). High-resolution genome-wide mapping of transposon integration in mammals. *Molecular and Cellular Biology*, *25*, 2085–2094.
44. Yergeau, D. A., Johnson Hamlet, M. R., Kuliyev, E., Zhu, H., Doherty, J. R., Archer, T. D., et al. (2009). Transgenesis in Xenopus using the Sleeping Beauty transposon system. *Developmental Dynamics*, *238*, 1727–1743.
45. Zayed, H., Izsvak, Z., Khare, D., Heinemann, U., & Ivics, Z. (2003). The DNA-bending protein HMGB1 is a cellular cofactor of Sleeping Beauty transposition. *Nucleic Acids Research*, *31*, 2313–2322.

Chapter 6
Insertional Mutagenesis in Hematopoietic Cells: Lessons Learned from Adverse Events in Clinical Gene Therapy Trials

Lars U. Müller, Michael D. Milsom, and David A. Williams

6.1 Introduction

From an early stage in the development of retroviral vectors for gene therapy applications, there has been a concern that recombinant vectors could elicit cellular transformation by altering expression of either cellular proto-oncogenes or tumor suppressor genes that are proximal to the genomic integration site (Fig. 6.1). This phenomenon, referred to as insertional mutagenesis, was characterized as a property of wild type gammaretroviral viruses (previously known as murine oncoretroviral viruses) which are the prototype virus from which the recombinant vectors used in the majority of the clinical trials described in this chapter were derived (reviewed in [178]). Wild type retroviruses fall into two categories with regards to their transforming properties: Slow transforming and acute transforming retroviruses. Acute transforming retroviruses contain viral versions of cellular proto-oncogenes such as Myb, Myc and Src and, as the name suggests, are able to induce cellular transformation within a short latency period [15, 62, 149]. Slow transforming retroviruses do not contain oncogenes but instead lead to transformation of the host cell via insertional mutagenesis. The fact that transformation by this route will likely require the accumulation of many proviral integrations within a given cell until a combination which can co-operate in transformation is achieved accounts for the longer latency period. Since cellular transformation was perceived to be a consequence of co-operating hits from multiple vector insertions, early pioneers in the gene therapy field focused upon developing vector systems which minimized the risk of generating a replication competent recombinant vector and which were unlikely to rescue dormant endogenous retroviral sequences present in the human genome. Such efforts involved the segregation of viral proteins essential for viral propagation (GAG, POL and ENV) onto packaging plasmids which were separate from the packaged recombinant viral genome. The identification of the gammaretroviral

D.A. Williams (✉)
Harvard Medical School, Children's Hospital Boston, Boston, MA 02115, USA
e-mail: dawilliams@childrens.harvard.edu

Lars U. Müller and Michael D. Milsom have equally contributed to this work.

A.J. Dupuy, D.A. Largaespada (eds.), *Insertional Mutagenesis Strategies in Cancer Genetics*, DOI 10.1007/978-1-4419-7656-7_6,

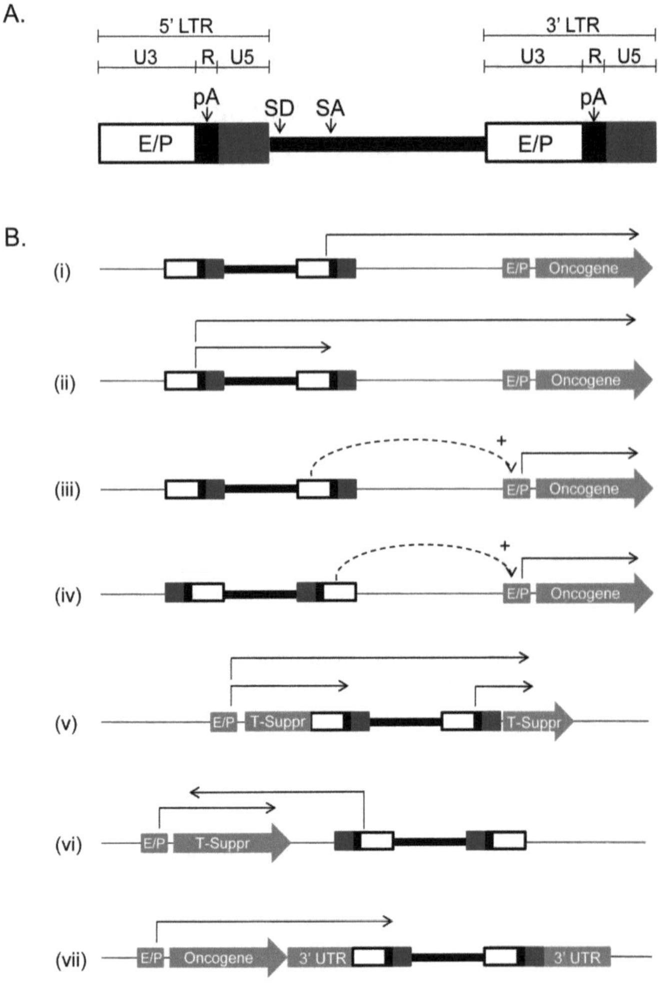

Fig. 6.1 Illustration of possible mechanisms through which recombinant gene therapy vectors may disrupt expression of a host cell gene. a Schematic representation of a recombinant retroviral vector highlighting the features which may contribute towards insertional mutagenesis. The vector schematic represents the integrated form of the provirus genome. The viral genome is flanked by two long terminal repeats which can be subdivided into the U3, R and U5 regions. The U3 region contains the transcriptional enhancer/promoter element which drives expression of the viral genome during packaging and which is required for expression of the transgene cassette. The R region contains the polyadenylation signal. Some vector configurations also incorporate a splice donor and splice acceptor site to enhance expression of the transgene cassette. Transgene cassettes may also contain cryptic splice donor/acceptor sites. **b** Molecular mechanisms through which an integrated provirus may promote cellular transformation by altering expression of flanking genomic loci. (i) Transcription initiated from the promoter in the viral 3′ LTR resulting in an mRNA containing a proto-oncogene located downstream of the integration site. (ii) Inefficient polyadenylation of the transcript initiated from the promoter in the viral 5′ LTR resulting an mRNA containing the recombinant viral transgene and a proto-oncogene located downstream of the integration site. Capture of the exons encoding the oncogenic protein may involve aberrant splicing.

packaging signal (Ψ) facilitated the deletion of this element in recombinant packaging constructs containing GAG, POL and ENV, thus eliminating the normal mechanism via which these coding sequences are incorporated into the viral particle [105]. In order to minimize the probability that segregated packaging constructs would recombine with either endogenous retroviral sequences in the target cell genome or with each other to form replication competent virus, packaging systems with minimal overlapping sequence homology were developed (reviewed in [112]). Stable cell lines to package gammaretroviral vectors, including the GP + envAm12 and PG13 lines, were created via stable transfection of packaging constructs into established cell lines [107, 112, 113]. Such packaging lines allowed the isolation and characterization of individual packaging clones following a subsequent stable transfection with a construct containing the recombinant retroviral genome. Although there are examples of replication competent virus being generated in packaging cell lines, individual clones can be determined as free of replication competent virus and can be subsequently expanded and used to produce large scale preparations of retroviral supernatant for use in clinical trials [36, 57, 140]. Such modifications in vector production technology made the risk of producing a supernatant which contained replication competent retrovirus almost negligible.

Having greatly reduced the likelihood that retroviral gene therapy vectors could generate replication competent virus, the risk of a recombinant vector being able to transform a cell via insertional mutagenesis was perceived to be very low [125]. The lack of efficacy in pre-clinical models using human hematopoietic stem and progenitor ($CD34^+$) cells somewhat shifted emphasis further away from the risks of insertional mutagenesis. How would it be possible to transform a human bone marrow stem cell with a recombinant retroviral vector if it was extremely difficult to even transduce human $CD34^+$ cells at a clinically relevant level with such a vector? Much of the energy in the field was therefore devoted to advancing stem cell transduction methodology and developing retroviral vectors which would

Fig. 6.1 (continued) (iii) and (iv) Enhancer effect from the viral U3-encoded enhancer element on an endogenous proto-oncogene promoter located either downstream (iii), or upstream (iv) of the integration site. (v) Integration within a tumor suppressor gene resulting in abrogation of gene expression either via premature termination of the transcript; the production of a fusion protein which may have no functional activity or dominant negative activity; or the introduction of a frame shift mutation resulting in premature termination of translation. The above may be mediated by an aberrant splicing event. (vi) Transcription initiated from the viral LTR resulting in the production of a mRNA which is complimentary to that of a tumor suppressor gene which is located downstream and inverted relative to the integration site. Such a complimentary mRNA may promote degradation of the tumor suppressor gene transcript or interfere with its translation (vii) Integration into the 3′ untranslated region of a proto-oncogene or tumor suppressor gene resulting in perturbed post-transcriptional regulation of gene expression either via destabilizing/stabilizing the mRNA; interfering with mRNA localization; or enhancing/repressing the initiation of translation. LTR, long terminal repeat; U5, unique 5′ LTR sequence; R, repeat LTR sequence; U3, unique 3′ LTR sequence; E/P, enhancer/promoter; pA, polyadenylation signal; SD, splice donor site; SA, splice acceptor site; +, enhancing effect upon transcription initiation; T-Suppr, tumor suppressor gene; 3′ UTR, 3′ untranslated region express transgene cassettes at levels which would be high enough to elicit a therapeutic benefit. Subsequent advances in vector design such as the optimization of LTR enhancer/promoter elements and viral leader sequences resulted in recombinant vectors which were able to mediate high level transgene expression in both primitive and mature hematopoietic cells [9, 40, 54, 75, 78, 104, 148]. Different viral envelopes were used to pseudotype recombinant particles in order to more efficiently target CD34 cells [60, 81, 84, 147, 153]. The development of improved in vitro growth media formulations incorporating novel cytokine cocktails achieved the dual aim of promoting hematopoietic stem cell division, which is required for transduction with gammaretroviral vectors, while minimizing stem cell loss in vitro via apoptosis or differentiation [13, 41, 46, 61, 85, 92, 133]. Where poly-cations such as polybrene had previously been used to enhance transduction frequencies by negating electrostatic charge repulsion between target cells and viral particles, the characterization of the CH296 fibronectin fragment (Retronectin) as a matrix upon which one could co-localize hematopoietic stem cells and viral particles was a significant advance in the quest to improve $CD34^+$ transduction frequencies [38, 70, 106, 127]. The development of antibody-based enrichment of the $CD34^+$ hematopoietic stem and progenitor cell compartment from human hematopoietic tissues using magnetic column purification provided a rapid, clinically applicable method to further enhance retroviral transduction by increasing the vector to target cell ratio [11, 12, 51, 175]. Taken together, these technological advances served as the platform for the first successful gene therapy trial in humans.

The correction of X-linked severe combined immunodeficiency (SCID-X1) via transplant of retroviral corrected human $CD34^+$ cells was reported in 2000 by Fischer and colleagues [34]. The success of this clinical trial finally realized the promise of retroviral-mediated gene therapy of human bone marrow $CD34^+$ cells, more than 15 years after the first demonstration that murine hematopoietic stem cells could be effectively transduced by recombinant retroviral vectors [189]. Moreover, a second independent gene therapy trial was able to recapitulate the correction of the SCID-X1 phenotype using a similar gene therapy protocol [58]. This validation of the potential effectiveness of gene therapy was particularly timely in the wake of the negative press that the gene therapy field received following the death of a patient in a phase 1 clinical trial for ornithine transcarbamylase deficiency, resulting from a systemic inflammatory response following administration of a recombinant adenovirus vector [146, 170]. However, soon after the publication detailing the efficacy of the Paris-based SCID-X1 trial, work emerged from the group of Baum and colleagues that would re-establish the importance of insertional mutagenesis as a significant risk factor in the retroviral-mediated genetic correction of hematopoietic cells, and which ultimately predicted the outcome of 5 out of 20 of the patients enrolled across both SCID-X1 gene therapy trials. Baum demonstrated for the first time that a replication incompetent retroviral vector backbone designed for gene therapy applications could cause cellular transformation via insertional mutagenesis in the context of a transplant model of transduced murine hematopoietic stem cells [101]. In this and a subsequent study, it was found that a single retroviral insertion in the vicinity of the ecotropic viral integration site 1 (Evi1) gene or the related

PR domain containing 16 (PRDM16) gene resulted in their overexpression and was sufficient to initiate a cascade of events resulting in leukemic transformation in vivo [101, 121]. Furthermore, a high copy number infection of murine bone marrow with recombinant retroviral vectors was able to facilitate combinatorial hits which caused leukemogenesis [120]. The pattern of cellular genes which combine to promote cellular transformation demonstrated a significant overlap with those that are disregulated in experiments which employed replication competent retrovirus vectors to provoke the development of leukemia [120]. Although murine hematopoietic stem cells likely represent a more readily transformed target than their human counterparts, these studies formally established the mutagenic potential of recombinant retroviral vectors intended for gene therapy applications.

Baum's group then made the seminal observation that at low copy number, retroviral-transduced murine hematopoietic stem cells are selectively expanded during transplant dependent upon proviral insertion site [93]. These non-malignant dominant clones are enriched for proviral integration sites in the locale of genes encoding signal transduction molecules and growth promoting genes [93, 94]. Analysis of mRNA expression levels in these clones revealed that the proviral insertion did indeed alter transcriptional regulation of genes proximal to the integration site and led to the hypothesis that this was a powerful method to identify pro-engraftment genes through positive selection. These observations were found to have direct translational relevance in a gene therapy trial for chronic granulomatous disease, where the non-malignant expansion of dominant retroviral transduced clones in two patients was found to correlate with insertional up-regulation of growth promoting genes [138].

In this chapter, we will discuss the results of key clinical gene therapy trials with a focus on retroviral-mediated insertional mutagenesis of human hematopoietic cells. We will also review how the serious adverse events in three of these trials have stimulated novel lines of research resulting in safety modification of gene therapy protocols.

6.2 Experience with X-Linked Severe Combined Immunodeficiency (SCID-X1)

Severe combined immunodeficiency (SCID) comprises a number of rare monogenic diseases with the common feature of a block in T-cell differentiation and impaired B-cell and natural killer (NK) cell immunity [52]. Studies of pattern of inheritance, immune function, and genotypes have led to the identification of at least 11 distinct SCID conditions. The most common variant of SCID results from the deficiency in expression or function of the common cytokine receptor γ chain, which is shared by the receptors for interleukin (IL)-2, IL-4, IL-7, IL-9, IL-15, and IL-21. This condition is inherited in a sex-linked fashion (X-linked SCID or SCID-X1) and accounts for 40–50% of all SCID cases [132, 174]. SCID-X1 is characterized by abnormal development or function of T, B and NK cells. Survival depends on the reconstitution of T-cell development and function by allogeneic bone marrow transplantation [5, 26]. If a genotypically matched family donor is available, hematopoietic stem cell transplantation (HSCT) confers greater than 80% chance of long term survival [5]. The absence of T- and NK cells in the patient allows for the engraftment of donor cells without preparative chemotherapy conditioning, and thus this is the treatment of choice with minimal toxicity. When a genotypically matched family member is not available, haploidentical donors (for example a parent) or closely matched unrelated donors are used, with varying preference from center to center, and survival of 64–78% has been reported [5, 14, 26, 63, 68]. These inferior outcomes may be attributed to the increased risk of graft rejection or graft versus host disease (GVHD), as well as the effects of T-cell depletion and immune suppression, causing slower immune reconstitution with increased risk of from infection [111, 130]. Haploidentical transplants rigorously depleted of T-cells, like genotypically related transplant, may be performed without preparative chemotherapy conditioning; however, B cell reconstitution is poor and the majority of patients require intravenous immunoglobulin replacement for life [111, 130]. Interestingly, spontaneous partial correction of severe T-cell immunodeficiencies, including SCID-X1and ADA-SCID, have previously been reported, suggesting a selective advantage of wild type T-cells over defective T-cells [23, 77, 171]. Therefore, two independent gene therapy trials, aimed at correcting the immunologic defect of SCID-X1 patients who lack a genotypically matched bone marrow donor have been undertaken [34, 58]. Thus far, a total of 20 patients have been treated on these studies. Despite minor technical differences in the two protocols, the basic design of both gene therapy trials is quite similar: The complete coding region of the human γ chain was cloned into a pMFG gammaretroviral vector regulated by the MLV long terminal repeat (LTR) sequences, which was used to infect bone marrow derived $CD34^+$ cells in vitro. The transduction occurred in the presence of early acting cytokines (stem cell factor, thrombopoietin, interleukin-3 (IL-3), and FMS-like tyrosine kinase 3 (FLT 3) ligand) and the CH296 human fragment of fibronectin. Cells were subsequently infused without prior conditioning or cytoreductive treatment. Minor differences between the two protocols include the uses of a three-fold higher concentration of IL-3 and 4% fetal cell serum in the French trial. Additionally, the French investigators utilized the amphotropic pseudotype, as compared to the use of gibbon-ape-leukemia-virus (GALV) envelope in the British trial [34, 58]. Results in both trials have been extremely encouraging.

In the French trial, 10 children under the age of 1 year were enrolled between 1999 and 2002 [33, 34]. Nine of ten infants developed normal numbers of T- and NK cells, with good immune function [64]. In seven of the nine patients who developed T-cells, T-cell counts reached normal levels within 3 months and have remained normal at the time of the last published follow up [32]. Protective levels of antibodies, including antibody production after immunization, were achieved and the prophylactic administration of intravenous immune globulin (IVIG) was discontinued [64]. At almost 8 years post gene therapy, these patients continue to retain a functional immune system, enabling them to live normally [32]. However, severe adverse events related to gene therapy has been reported in four patients in the French trial, occurring 31–68 months post gene therapy [32, 65, 66]. In these patients, untoward effects of viral integration into the genome resulted in clonal T-cell proliferation, leading to the death of one of the four affected patients. Much research has subsequently been directed at elucidating the mechanism responsible for these adverse events. It has now been clearly demonstrated that retroviral integration in the proximity of proto-oncogenes, particularly the LIM domain only 2 (LMO2) promoter, was involved in leukemogenesis in three patients of the French patients and one British patient [32, 80]. An integration of the unaltered γ chain-encoding viral vector on chromosome 11q13, near the first exon of the LMO2 gene, led to the unregulated transcription of LMO2, giving rise to a T-cell acute lymphoblastic leukemia (T-ALL)-like lymphoproliferation in the initial two patients [65]. LMO2 is a master regulator of human hematopoiesis, involved in stem cell growth and is not normally expressed in T-cells. However, LMO2 activation has been implicated in some cases of human T-cell leukemia [193]. In addition, LMO2 transgenic mice have been shown to develop T-ALL within 10 months [55, 129]. It is increasingly clear that retroviral vectors may "turn-on" cellular proto-oncogenes adjacent to their integration site in the genome. The strong promoter/enhancer activity of the retroviral LTR element shows particular propensity to the up regulation of genes neighboring the integration site [93, 101, 120]. Multiple studies now indicate that gammaretroviral vectors such as the vectors used in the two SCID-X1 trials preferentially integrate into the 5′ end of genes, near the transcription start unit [192]. In addition, gammaretroviral vectors have been shown to integrate in or near a number of proto-oncogenes that are actively expressed in human $CD34^+$ cells. When human $CD34^+$ cells were transduced with retroviral vectors in vitro, 21% of retroviral integrations occurred at recurrent insertion sites ("hot spots"), which were highly enriched for proto-oncogenes and growth-controlling genes [31]. A recent series of papers investigating the vector integration sites in both SCID-X1 trials and the Italian ADA-SCID trial, observed a greater than random frequency of vector integrations near the transcription start site of genes that are active in hematopoietic stem cells [1, 45, 163]. Interestingly, in the SCID-X1 trial a skewing of vector integration site distribution in vivo was noted. Compared to retroviral integration sites (RIS) recovered from transduced $CD34^+$ cells, RIS recovered from T-cells in vivo 9–30 months after transplantation, showed an overrepresentation of RIS within or near genes encoding proteins with kinase activity, transferase activity, or proteins involved in phosphorous metabolism. This skewing of RIS in vivo suggests a selection of T-cells as a result of viral integration in certain growth and survival promoting genes [45, 163]. Strikingly, in contrast to 5 cases of insertional mutagenesis in the two SCID-X1 trials, no adverse events have been reported in the 10 patients treated in the Italian ADA-SCID trial, despite a similar RIS pattern observed in this patient group [1]. This observation has led to the proposal of a "disease effect" contributing to oncogenesis of γ chain gene therapy. Woods et al. demonstrated lentiviral transduction of γc–/– mice with vectors containing the human common γ

chain (cγ) or an inert control gene at very high viral doses. They observed the induction of T-cell malignancies in a third of the animals receiving cγ transduced cells, but not in the control groups [191]. The merit of this very limited study was subsequently challenged, primarily on the basis of a viral dose much higher than that used in clinical scenarios and incomplete data on the pathogenesis of the malignancies, in particular as it relates to the downstream activation level of a key signaling target of the common γ chain, JAK3, in the tumors [142, 177]. In addition, the lentiviral vector used in this study incorporated a hybrid promoter/enhancer element which is extremely powerful and likely to possess a greater trans-activating potential than promoter/enhancer elements that would be considered for clinical gene therapy use [191].

The experience of the 10 patients treated in the British SCID-X1 trial has recently been updated [80]. Previously, the scientific community was perplexed at the lack of adverse events in this trial, despite a similar approach to that used in the French trial. Even in the absence of a statistically significant difference in the occurrence of insertional mutagenesis induced leukemia (now 4/10 in the Fresh trial and 1/10 in the British trial), questions regarding the viral pseudotype (amphotropic versus GALV), the use of fetal cell serum, and the dosing if IL-3 were raised. It was hypothesized that these factors may have contributed to the more rapid kinetics of T-cell reconstitution that were observed in patients in the Paris trial [34, 58]. It now appears that these minor differences are unlikely to contribute to a significant difference in outcome. The shorter observation period in the British trial and small sample size are more likely to account for this difference. Howe et al. are now reporting the occurrence of a T-ALL, promoted by insertional mutagenesis in one of the 10 British SCID-X1 patients. In this patient, the integration of the vector 35 kb upstream of the LMO2 locus cooperated with secondary genetic aberrations, including a gain of function mutation of NOTCH1, a deletion at the CDKN2A tumor suppressor gene locus, and a translocation of the T-cell receptor-β region, to give rise to T-ALL [80]. The patient had been treated with gene therapy for molecularly confirmed SCID-X1 at 13 months of age and initially experience satisfactory immunologic recovery. IVIG replacement was temporarily suspended at 13 months of age but recommenced at 16 months due to persistently low immunoglobulin levels. The patient was diagnosed with T-ALL at 24 months of age and achieved a complete remission with standard T-ALL chemotherapy.

In summary, thus far 5 of 20 patients treated with gene therapy for SCID-X1 have encountered a live-threatening severe adverse event, thought to be triggered by retroviral activation of LMO2 in four patients. Four patients were salvaged with chemotherapy and one patient succumbed to the disease following an unsuccessful allogeneic bone marrow transplantation. The risk and benefits of gene therapy for SCID need to be carefully reassessed. Clearly, the use of MLV-based retroviral vectors with LTR promoter enhancer elements is viewed as contraindicated in this disease by most investigators in the field. The continued development of safety-enhanced vectors and the validation of these vectors in clinically relevant systems has emerged as a major priority in the field.

6.3 Experience with Adenosine Deaminase Deficiency Severe Combined Immunodeficiency

Adenosine deaminase (ADA) is a housekeeping enzyme of the purine metabolic pathway, expressed in all tissues of the body [73]. Deficiency of this enzyme leads to a build-up of toxic metabolites with detrimental systemic effects, including neuro-developmental deficiencies, sensori-neuronal deafness, and skeletal abnormalities. Importantly, ADA deficiency causes abnormal T, B, and NK cell development, resulting in severe-combined immunodeficiency. As is the case with the more common SCID-X1, untreated patients generally succumb to severe opportunistic infections in the first year of live. Treatment strategies employed to manage affected patients include allogeneic stem cell transplantation (SCT), enzyme replacement therapy, and more recently gene therapy [19]. Allogeneic SCT from an HLA-matched family donor offers good immunological and biochemical correction with 73% survival. However, outcomes following mismatched and haploidentical transplants are less impressive [5]. Likewise, the exogenous replacement of ADA, administered in a polyethylene glycol (PEG) conjugate by intra-muscular injection on a weekly or twice weekly schedule, will result in systemic detoxification and immune reconstitution. In the long term, however, about half of the patients receiving PEG-ADA replacement continue to require IVIG infusions and some patients show a decline in T-cell numbers over time. A number of gene therapy trials for ADA deficiency were initiated in the early 1990s, targeting retroviral gene transfer into various cell types, including peripheral blood lymphocytes, umbilical cord blood, bone marrow, and CD34^{+} selected stem cells [16, 20, 79, 90]. These early studies failed to produce clear efficacy. By contrast, more recent studies introduced key modifications to the gene therapy protocol, including the use of a reduced intensity myelosuppressive conditioning regimen and the withdrawal of PEG-ADA replacement [2, 59]. The Milan-based group of Aiuti and colleagues has thus far enrolled ten children. Patients were conditioned with 4 mg/kg of Busulfan prior to the infusion of transduced cells. The mean age at the time of gene therapy was 2.2 years. All children on this trial are healthy and thriving, with the longest published follow-up now over 64 months [1, 3, 19]. Gene therapy has resulted in a substantial increase of lymphocyte counts and normalization of T-cell function. Similarly, four patients have been treated in London [179]. One patient that has been reported in detail had been treated with PEG-ADA for 3 years but showed a gradual decline in T-cell numbers, despite effective metabolic correction. As a matched bone marrow donor was not available, the patient was enrolled on the ADA-SCID gene therapy trial. PEG-ADA replacement was stopped 1 month prior to gene therapy and the patient was conditioned with a single dose of 140 mg/m^{2} of Melphalan prior to the infusion of the transduced bone marrow CD34^{+} cells [59]. At the time of the last published follow up, the patient was 2 years out from gene therapy, clinically well and off prophylactic antibiotic therapy. An increase in T-cell numbers and normalization of the proliferative response have been noted [19, 59]. Importantly, no adverse events have occurred thus far in the patients treated for ADA-SCID at these two centers. Aiuti et al. recently published a comprehensive genome wide analysis of retroviral integrations sites (RIS) of five patients treated in Milan [1]. This paper analyzed the RIS patterns in $CD34^+$ cells prior to infusion as well as RIS in vivo, up to 47 months post gene therapy. As anticipated, a non-random pro-viral integration pattern, favoring transcription start sites (TSS) and gene dense regions, was observed in the pre-transplant cells. RIS observed in vivo in T-cells were additionally enriched for TSS, suggesting the occurrence of in vivo selection. More recently, Aiuti and colleagues have demonstrated that cellular genes in the proximity of the proviral integration site are subject to moderate disregulation in gene modified T-cell clones isolated from patients [30]. However, in contrast to the SCID-X1 trial, no in vivo skewing toward RIS in genes affecting survival, cell cycling, signal transduction, or proliferation were observed, making a clonal dominance effect appear less likely. Interestingly, only one RIS was detected at the MDS-EVI1 locus and became undetectable at later time points. This is in contrast to the clonal dominance of MDS-EVI1 integration sites observed in the X-CDG trial [138]. Additionally, an overrepresentation of RIS was noted in the proximity of the CCND2 and LMO2 gene with a total of 5 of 523 RIS recovered in vivo. Notably, the CCND2 insertions were detected only in the first 2 years of follow up and not subsequently. LMO2 insertions were also overrepresented in the pre-transplant $CD34^+$ samples, highlighting the fact that the LMO2 gene is a hot spot for retroviral integration in human $CD34^+$ cells [1]. The lack of in vivo expansion of clones carrying LMO2 RIS indicates that this integration site may not be sufficient to mediate clonal dominance and leukemic transformation. Rather, additional cooperating mutations or insertions are required for malignant transformation. The lack of malignant transformation in two ADA-SCID trials may point to the role of the genetic background or the role of the therapeutic transgene introduced into human $CD34^+$ cells. However, the patient cohort remains relatively small and follow up is still short term. Overall, the genotoxicity profile in these two ADA-SCID trials has been sufficiently favorable to continue to recommend this experimental therapy to patients and families lacking a perfectly matched sibling donor.

6.4 Experience with Chronic Granulomatous Disease

Chronic granulomatous disease (CGD) is an inherited disorder of phagocyte dysfunction, characterized by often life threatening invasive fungal and bacterial infections and by granuloma formation in vital organs. CGD results from a mutation in one of four subunits of the nicotinamide adenine dinucleotide phosphate (NADPH) oxidase of phagocytes. The inability to form microbiocidal oxygen species renders the phagocytes unable to fight invasive infections [165]. Almost 70% of CGD cases result from defects in the X-linked gene encoding $gp91^{phox}$ (X-CGD). With conventional therapy, including lifelong antimicrobial prophylaxis and interferon-gamma therapy, the yearly mortality rate of X-CGD remains at 5% [190]. Bone marrow transplantation is curative for patients with a perfectly matched sibling donor but remains risky in patients with active infections. Unrelated donor transplantations are not routinely recommended [166]. Thus, the development of a gene therapy approach that utilizes autologous HSC provides an important therapeutic advance for this patient group. In previous clinical gene therapy trials conducted without myeloreductive conditioning, the engraftment level of gene modified cells remained low [103]. In 2002, the German group of Grez and colleagues in Frankfurt, Germany, initiated a gene therapy trial of X-CGD. The initial patients received a mild immunosuppressive preparative regimen and failed to engraft significant numbers of gene modified cells. However, 2 years later, low-dose busulfan – modeled on the successful gene therapy trial for ADA-SCID [2] – was incorporated into the preparative regimen and an additional six patients have been treated thus far [137]. This group of patients has been followed with unprecedented sophistication by the prospective monitoring of integration sites that mark each hematopoietic cell prior to transplantation and then allow the tracking of these cells in vivo [138]. The initial two patients treated were 26 and 25 years old, respectively. Both subjects carried the diagnosis of X-CGD and had failed to clear invasive infections, including a Staphylococcus aureus liver abscess and pulmonary aspergillosis, with medical treatment. Thus, autologous peripheral blood $CD34^{+}$ cells were mobilized with G-CSF and collected. Gene transfer was performed utilizing a gammaretroviral vector SF71gp91^{phox}. This vector, containing the spleen focus-forming virus long terminal repeat (LTR) elements was chosen for its ability to achieve high expression levels in transduced hematopoietic stem cells [76]. The in vitro transduction rates in the two patients were 45 and 39.5%, respectively, with a pro-viral copy number of 2.6 and 1.5 per transduced cell. Pro-viral integration occurred preferentially in gene-coding regions (47–52%) and was highly skewed towards the 5 kb sequence surrounding the transcription start sites. Moreover, the clonal distribution pattern was not stable over time. Rather, starting 5 months after therapy, a less diverse integration pattern emerged, indicating the appearance of dominant clones. Clinically, following a period of cytopenia after the conditioning and cell infusion, the initial engraftment rates detected in the peripheral blood were 12–13%. Significant improvement in the previously refractory infections was noted 50–60 days after therapy. Surprisingly, a gradual increase in the number of gene-corrected cells up to 50–60% of all peripheral blood cells was observed, starting around day 150 post transplant. This coincided with increased oxidase activity and occurred in the absence of altered blood counts. These events were accompanied by a selective outgrowth of progenitors carrying vector insertions that activated one of three oncogenes, *PRDM16, SETBP1*, and most notably *MDS-EVI1*. While all three genes are well-known cancer-associated genes, most clonal outgrowths were exhausted after a few months with the exception of MDS-EVI1, which increased to 67–90% in both patients approximately 1 year post cell infusion. Of note, the dominant MDS-EVI1 clones initially did not transgress the boundaries of the normal myeloid pool, as these cells remained cytokine-dependent in vitro and failed to engraft in immunocompromised mice, suggesting their benign nature [138]. Thus, the expansion of gp91^{phox+} cells clearly provid therapeutic benefit during the initial phase. More recent follow up on these two study patients has been provided in abstract form [139]. Indeed, while gene marking remained high in both patients, down regulation of trans gene expression was noted as a result of CpG methylation in the viral LTR promoter. As a consequence, gp91phox expression was suppressed but the capacity of the LTR encoded enhancer to transactivate nearby genes remained intact [139]. One patient died 2.5 years post therapy of severe sepsis [4]. The second patient developed monosomy 7 and myelodysplastic syndrome (MDS) and died following an unsuccessful unrelated donor BMT (personal communication, Manuel Grez). Of note, the *EVI*1 locus has previously been identified as a common target of retroviral oncogenesis [27, 101]. *EVI*1, which is not detected in normal hematopoietic cells, has been associated with myeloid leukemia and MDS [8, 150]. The constitutive overexpression of Evi1 in mouse bone marrow cells has been shown to induce MDS in mice [28]. Despite these molecular events, the infusion of gene corrected $CD34^+$ cells was highly effective with regard to clearing refractory pyogenic infections [138], raising the possibility of using gene therapy to bridge patients with refractory pyogenic infections into eligibility for allogeneic HSCT.

6.5 Preliminary Data from Other Trials

In addition to the five gene therapy clinical trials that we have already discussed, there is preliminary data from at least three other clinical trials in which significant marking of reconstituting $CD34^+$ cells has been achieved.

6.5.1 Wiskott-Aldrich Syndrome (WAS)

WAS is an X-linked immunodeficiency which is caused by inactivating mutations in the WAS protein (WASP). WASP plays a regulatory role in cell signaling and cytoskeletal reorganization in hematopoietic cells [22]. The disease is fatal and is characterized by severe combined immunodeficiency, thrombocytopenia, elevated frequency of tumor formation, eczma, and other autoimmune manifestations [134]. The only currently available curative therapy for WAS is bone marrow transplant but, as with the other primary immunodeficiencies, the availability of suitably matched donors is limiting [135]. A clinical trial for the genetic correction of WAS via retroviral delivery of the WASP cDNA into autologous $CD34^+$ cells is currently underway [25]. A combination of a relatively high cell dose (8–7 $\times$ 10^6 $CD34^+$/kg body weight) and good transduction efficiency led to gene marking across both myeloid and lymphoid lineages. A marked clinical benefit from gene therapy has been reported in one of the patients. Both patients are being closely monitored for evidence of clonal imbalance, as determined by molecular analysis of proviral integration site pattern. To date, there has been no reported incidence of clonal expansion of transduced hematopoietic cells, nor has there been any evidence of morphologic or cytogenetic abnormalities in bone marrow, although the period of follow up is short. Of particular note is the fact that the architecture of the vector backbone used in this trial is similar to that employed in both the CGD and the London-based ADA-SCID trials described above. That is, the gammaretroviral vector has an intact LTR which contains the enhancer/promoter from the spleen focus forming virus.

6.5.2 X-Linked Adrenoleukodystrophy (ALD)

ALD in children is a fatal neurodegenerative disease which results from progressive neural demyelination within the brain [83, 128]. The defective gene which is responsible for the phenotype, ATP-binding cassette D1 (ABCD1), encodes a transmembrane transport protein which is responsible for the shuttling of fatty acids into peroxisomes where they are subsequently degraded [87]. ALD is characterized by an accumulation of very long chain fatty acids, although the exact pathophysiology of the disease is unknown. Allogeneic bone marrow transplant has been found to be therapeutic for cerebral demyelination, presumably due to the infiltration of donor-derived microglia cells into the brain [168]. However, because of the progressive nature of the demyelination and the time required to generate mature microglia from transplanted hematopoietic stem cells, bone marrow transplant is most effective as soon after development of demyelination evident by imaging methods as possible [168]. Because of the time constraints imposed upon finding a suitable HLA-matched donor, the genetic correction of autologous bone marrow $CD34^+$ cells is an attractive experimental therapeutic option. A phase 1 gene therapy trial using a recombinant HIV-1 based lentiviral vector to deliver the ABCD1 cDNA into $CD34^+$ cells is currently underway in Paris [35]. In the three patients that have been enrolled to date, a transduction efficiency of 30–50% in $CD34^+$ cells was achieved with mean vector copy numbers of 0.6–0.7 per cell. Patients were pre conditioned with cyclophosphamide and busulfan and between 9 and 23% multilineage gene marked chimerism has been reported in a follow up period that extends up to 16 months. In the two patients that have been followed for 16 months post-transplant, hematopoietic stem cell gene therapy resulted in neurologic stabilization which are similar to those achieved with allogeneic transplant. This trial is particularly noteworthy in that it is the first trial to report reconstitution of bone marrow cells which have been transduced with a recombinant lentiviral vector. Molecular characterization of lentiviral integration sites in engrafting cells indicated that reconstitution was polyclonal but detailed analysis of genomic loci targeted by this vector has yet to be reported.

6.5.3 β Thalassemia

A second clinical trial which has employed a lentiviral vector is taking place in Paris for the correction of β-thalassemia. Thalassemia's result from mutations which attenuate the expression of either the α or β globin chains which compromises hemoglobin synthesis and thus causes inefficient erythropoiesis [136, 187]. As adult hemoglobin consists of a tetramer of two α and two β chains, inherited mutations at the β globin locus cause a mismatch in the ratio of these two chains and thus prevents the correct assembly of the hemoglobin molecule. Clinically, this may result in transfusion dependent anemia which in turn can promote the serious side effect of iron overload. In general, disease severity correlates with the degree to which mutations inhibit β globin expression. However, there are other genetic loci that can modulate the disease phenotype, for example by inducing adult expression of the fetal γ globin gene which can be efficiently incorporated into functional hemoglobin in place of the β globin chain [154, 187]. Gene therapy of β-thalassemia is complicated by the requirement for an exact stoichiometry of α and β globin chains in order to facilitate efficient assembly of hemoglobin. Thus, an effective gene therapy vector must be able to facilitate high level expression of β-globin in the range of that mediated by the normal endogenous gene in an erythroid specific context. Lineage specific expression of the β globin chain is particularly important in the context of genetic modification of HSC in order to prevent high level expression of β globin in other hematopoietic lineages.

The lentiviral vector used in this trial comprises a SIN configuration with elements from the β globin locus control region (LCR) driving expression of the β globin cDNA which has been mutated to enhance β globin chain stability. Additionally, the vector's expression cassette is flanked by chromatin insulator elements which putatively function to both prevent silencing of the transgene expression cassette by inhibitory chromatin structure surrounding the integration site, and to prevent the vector-encoded enhancer from modulating expression of endogenous genes near to the insertion site (6.4 Modification of vector architecture to minimize insertional mutagenesis). An interim report on the findings of the clinical trial describe that in 2007, a single patient suffering from severe transfusion-dependent β thalassemia major received $CD34^+$ cells transduced with the lentiviral vector described above *[Philippe Leboulch, presentation to the 2009 Annual Meeting of the American Society of Gene Therapy; Agence Française de Sécurité Sanitaire des Produits de Santé report]*. There was a clear demonstration of clinical efficacy as the patient has been transfusion independent for 15 months. Of note, post-transplant molecular analysis revealed a mild clonal skewing comprising less than 5% of peripheral blood cells. The clone, which harbors a pro-viral integration within the high mobility group A2 proteins (HMGA2) gene locus, is reported to be stable. Insertion of the provirus into this locus resulted in the production of a 3′ truncated mRNA resulting from the introduction of a cryptic splice acceptor, present in the vector insulator element, into intron 3 of the gene. This truncated mRNA, comprising exons 1–3, was expressed at elevated levels due to a loss of negative post-transcriptional regulation by Let-7 miRNAs, since the miRNA target sequence in exon 5 was lost. HMGA2 has been found to be mutated in chromosomal translocations primarily in benign tumors and less often in malignant tumors [37]. Although the clinical implications of this clonal outgrowth are unclear, this event clearly demonstrates that lentiviral vectors can contribute to insertional mutagenesis albeit in this case via modulation of post-transcriptional regulation of gene expression.

6.6 Lessons Learned and the Way Forward

In considering the lessons learned from the clinical trials described in this chapter, one must recognize that these trials represent pioneering work which demonstrates that using current technology and across a range of inherited disorders, hematopoietic stem cells can be corrected by gene transfer and subsequently transplanted back into patients. Following autologous transplantation, these cells have elicited therapeutic effects. Despite the serious adverse events that have been reported in three of these trials, it is important to weigh the risk versus benefit ratio for each individual trial in the context of the conventional therapeutic options which are currently available. Gene therapy certainly holds a significant advantage over conventional allogeneic transplant in terms of facilitating a therapeutic autologous transplant. That is, the use of autologous cells reduces the likelihood of transplant-related morbidity and mortality from the side effects of ablative conditioning regimens and from graft versus host disease and the ensuing immunosuppressive regimens. Having said this, the onus is clearly upon the gene therapy community to make every effort to modify current gene therapy protocols in order to minimize the likelihood of adverse events occurring in future trials. Indeed, since the first report of insertional mutagenesis in the SCID-X1 trial was published in 2003, the efforts of the field have been focused upon defining the mechanisms through which gene therapy vectors elicit insertional mutagenesis and developing new technologies to overcome this problem. In this section, we will summarize the significant advances that have been made towards these goals.

6.6.1 Advances in the Analysis of Integration Site Preference

To facilitate the characterization of retroviral integration sites from clinical and pre-clinical samples, the group of von Kalle and colleagues has pioneered the development of two PCR-based strategies for amplifying and isolating the genomic DNA flanking proviral insertion sites. Both extended primer tag selection ligation-mediated (EPTS LM) PCR and linear amplification-mediated (LAM) PCR allow the isolation of discrete genomic integration sites from highly complex DNA samples [160–162]. Because of the superior performance of these techniques over previous technologies in terms of allowing robust recovery of integration sites from a modest amount of DNA, EPTS LM and LAM PCR have been used by the vast majority of researchers to interrogate samples. More recently, these technologies have been combined with high throughput sequencing techniques in order to facilitate the identification of thousands of integration sites within a single sequencing reaction [185]. In the wake of the Paris trial, PCR-based integration site analysis has been invaluable in establishing the clonality of gene modified grafts and measuring the development of clonal imbalance in both pre-clinical and clinical samples. Integration site analysis has also been extensively used to investigate the integration site preferences of different gene therapy vectors in order to determine whether some systems were inherently safer than others.

Wild type and recombinant retroviral vectors (including alpha, gamma, spuma and lenti) integrate into the host genome in a semi-random manner and demonstrate insertion site biases which are dependent upon the accessibility of the insertion site in the target cell and variations in the viral integrase enzyme which depend upon retroviral genus [42, 96]. Gammaretroviruses such as murine leukemia virus (MLV) have been shown to exert a clear preference for integration in the region immediately surrounding the TSS of actively transcribed genes [95, 118, 183, 192]. While lentiviral vectors also demonstrate a preference to integrate within the loci of actively transcribed genes, their integration profile favors sites which are downstream of the TSS represented within the body of the primary transcript [10, 31, 44, 67, 72, 118, 192]. Using viral chimeras, it has been shown that incorporation of MLV integrase into an HIV-1 based vector alters the integration pattern of the lentivirus to more closely resemble that associated with a gammaretroviral vector [97]. It is possible that this phenomenon results from the binding of the MLV integrase with cell type specific transcription factors resulting in the recruitment of the pre-integration complex to the promoter/enhancer region of actively transcribed genes [50]. This work clearly demonstrates that gammaretroviral and lentiviral vectors have developed distinct mechanisms of integrase-dependent integration which may have an impact upon the mutagenic potential of recombinant retroviral vectors. If one considers the possibility of integrating vectors upregulating oncogene expression via either readthrough transcription or enhancer effects on the endogenous promoter, then gammaretroviral vectors could be considered as potentially more mutagenic than lentiviral vectors in this context, due to their preference for integration near the TSS (Fig. 6.1). Conversely, preferential integration within the body of the primary transcript may result in lentiviral vectors having a higher probability of interrupting tumor suppressor gene expression (Fig. 6.1). Progress has been made in the development of model systems to functionally evaluate the relative mutagenic potential of different vector systems (6.2 Novel model systems to characterize insertional mutagenesis). However, the model systems developed to date have a clear preference to detect mutagenesis mediated via upregulation of oncogene transcription. It is not clear whether this is a reflection of tumor suppressor gene inactivation being inconsequential as a mechanism of insertional mutagenesis, or is a result of bias within the model system. Clearly, the preliminary results from the β thalassemia trial described above (5.3 β thalassemia) demonstrate that lentiviral vectors may mediate insertional mutagenesis via alternate mechanisms.

Other related retroviral vector systems have also been shown to have an integration pattern which is distinct from gammaretroviral vectors and as such may represent a safer vector configuration. Recombinant foamy virus vectors do not preferentially integrate within genes and their integration pattern does not significantly correlate with actively transcribed genes [180]. Likewise, avian sarcoma leukosis virus vectors do not favor gene rich regions or TSS as preferred integration sites [82]. However, these novel vectors systems have not been as well characterized as gammaretroviral or lentiviral vectors with regards to safety and efficacy, therefore they unlikely to be translated to clinical use in the near future.

6.6.2 Novel Model Systems to Characterize Insertional Mutagenesis

A number of different model systems have been devised to assess the relative mutagenic potential of different vector systems or protocols (reviewed in [131]). These include in vitro assays using cell lines or primary cells with gene activation, establishment of growth factor independence, or immortalization as a readout following retroviral transduction [18, 21, 49, 108, 120, 151]. Additionally, a number of murine in vivo assays have been developed which examine the onset of clonal dominance or transformation as a result of retroviral transduction. These in vivo models include the use of high copy number transduction; leukemia prone knockout mice; or disease-specific knockout/transgenic mice [93, 119, 123, 164, 169]. Despite the variations in experimental design encompassed by these models, a number of the findings appear to be consistent across several studies. First, gammaretroviral vectors which harbor an LTR-containing promoter/enhancer readily contribute to clonal dominance and cellular transformation via oncogene upregulation [93, 94, 101, 119, 120, 123, 151]. Second, lentiviral vectors which harbor an LTR-containing promoter/enhancer are also able to mediate insertional mutagenesis via similar molecular mechanisms to gammaretroviral vectors albeit with a lower incidence of transformation [18, 108, 124]. Third, a self-inactivating (SIN) configuration incorporating an internal promoter is less mutagenic for both gammaretroviral and lentiviral vectors [120, 122, 124]. Fourth, the use of a weaker cellular internal promoter dramatically reduces the genotoxic potential of both gammaretroviral and lentiviral vectors [122, 124, 196]. The one notable exception to this observation is a recent study that demonstrated immortalization of primary murine bone marrow cells using a high copy number infection with a SIN gammaretroviral vector containing the human cellular phosphoglycerate kinase internal promoter [21]. Fifth, the propensity of retroviral vectors to upregulate oncogene expression is directly linked to the transcriptional enhancer element contained within the vector [39, 108, 122, 196]. Sixth, given a comparable vector configuration and promoter/enhancer, lentiviral vectors may have a slightly reduced ability to upregulate oncogene expression compared to gammaretroviral vectors [122, 124]. Seventh, the majority of these model systems highly favors insertional mutagenesis via abnormal induction of oncogene expression and may not fully evaluate the transforming potential of recombinant retroviral vectors. As discussed previously, this last point could indicate that tumor suppressor gene inactivation is not a major mechanism through which retroviral-mediated insertional mutagenesis operates, or it could be attributed to the lack of sensitivity of these assays to detect transformation via downregulation of tumor suppressor gene inactivation. Certainly when using cells from highly inbred mouse strains where the majority of genetic loci will comprise the homozygous state, one can hypothesize that the tandem inactivation of both alleles of a tumor suppressor gene within the same cell is highly unlikely. Indeed, Copeland and colleagues demonstrated that in the mouse model, a genetic background which promoted an increased frequency of sister chromatid exchanges was required to detect tumor suppressor gene inactivation by insertional mutagenesis [173]. Notably, in a study which used a real time PCR-based approach to examine genotoxicity as a function of altered gene expression independent of biological effect, lentiviral vectors were found to mediate downregulation of gene expression with an equivalent frequency to which they were able to upregulate gene expression [108]. In contrast, gammaretroviral vectors were found to not contribute to decreased gene expression.

Gene therapy protocols have also been assessed using large animal transplant studies incorporating outbred dogs and non-human primates. For the most part, these studies have provided little evidence of transformation or clonal dominance resulting from retroviral insertional mutagenesis [24, 86]. Dunbar and colleagues have observed an overrepresentation of integrations near to the MDS1/EVI1 locus in long-term primate repopulating cells which had been transduced with gammaretroviral vectors but not in equivalent cells transduced with simian immunodeficiency virus-based lentiviral vectors [29, 88]. However, this was not associated with any evidence of clonal outgrowth or leukemia. Conversely, a comparison of integration sites between an HIV-1 based lentiviral vector and an MLV-based gammaretroviral vector in repopulating cells from pigtailed macaques and baboons revealed that both vector systems demonstrated a propensity to integrate near to oncogenes [10]. Again, there were no documented adverse effects arising as a result of retroviral transduction. Other than a study which detailed the induction of T-cell lymphomas in rhesus macaques which arose from $CD34^+$ transduction by a contaminating replication competent recombinant MMLV vector, there have been two documented cases of leukemia arising as a result of retroviral insertional mutagenesis in large animal studies [47, 182]. The first was also reported by the group of Dunbar and colleagues and relates to a study which examined the delivery of a mutated version of the drug resistance gene dihydrofolate reductase into $CD34^+$ cells from a rhesus macaque using an LTR-driven gammaretroviral vector. Five years after transplant with transduced $CD34^+$ cells, the animal developed a fatal AML resulting from a transduced clone harboring two retroviral insertion sites [167]. One of these insertion sites was within the *BCL2-A1* anti-apoptotic gene suggesting that this event may have initiated leukemic transformation. Further interpretation of this study is complicated by the fact that this animal received an anti-folate-based chemotherapeutic regimen following hematopoietic reconstitution which could have contributed to evolution of leukemia through provoking cell stress and DNA damage. The second large animal study to report an incidence of leukemia arising from insertional mutagenesis details the use of an LTR-driven gammaretrovirus to overexpress the homeobox transcription factor HOXB4 in both canine and non-human primate repopulating cells [195]. HOXB4 has previously been shown to perturb hematopoietic differentiation in both murine and human HSC/P but does not lead to leukemic transformation in murine transplant/xenotransplant studies [116, 155, 159]. Although the leukemic cells were shown to be dependent upon HOXB4 overexpression to elicit a block in differentiation, the leukemic clones isolated from the three diseased animals all had retroviral insertion sites which activated expression of oncogenes including *PRDM16* and C-MYB. Of particular note, a leukemic clone isolated from a macaque harbored two insertions within introns of the tumor suppressor gene single-stranded DNA binding protein 2 (*SSBP2*) which resulted in its downregulation. Clearly, large animal studies appear to be good models to study the long term effects of retroviral genotoxicity on transplanted HSC. Unfortunately, the cost of performing these studies prohibits the analysis of retroviral-mediated mutagenesis that has been performed in murine models due to the limitation in the number of animals that can be included in each study. Ultimately, it would be ideal if an assay could be developed to assess retroviral genotoxicity in the context of human CD34^{+} cells.

6.6.3 Disease Specific Effects Upon Insertional Mutagenesis

Several model systems have been developed to examine potential disease specific effects upon the incidence of adverse effects from insertional mutagenesis. Given the controversy surrounding the potential role of cγ overexpression in the pathogenesis of the leukemia observed in the SCID-X1 trial, significant effort has focused upon determining whether there is indeed a disease-specific component which was required for the high incidence of leukemia observed in the British and French trials [142, 177, 191]. Shou at al., describe the derivation of a tumor prone mouse in which both the Arf tumor suppressor gene and the cγ gene were knocked out [169]. Using this model, a high rate of transformation was observed upon retroviral-mediated gene therapy which was dependent upon both insertional activation of proto-oncogenes and upon the SCID-X1 background. It was noted that the cγ knock-out was associated with an increased frequency of hematopoietic progenitors which the authors hypothesize may act as a source of cells which are amenable to transformation. In a separate study, the large scale identification of integration sites within leukemic clones generated using replication-competent retrovirus identified that cells which harbored integrations near the *LMO2* locus were more likely to also contain insertion sites near the *cγ* gene [43]. Although the authors contend that the combined mutagenesis of the *LMO2* and *cγ* loci is insufficient to directly promote leukemogenesis and requires other co-operating oncogenic hits, they conclude that the overrepresentation of leukemic clones which contain integration sites near to both *LMO2* and *cγ* suggests a synergistic effect in transformation. In direct contrast, Pike-Overzet et al. demonstrate that LMO2 overexpression is sufficient to block the in vitro differentiation of T-cells from transduced human CD34^{+} cells, but that cγ overexpression has no effect upon this process [143]. Instead, these authors propose that repair of cγ expression is required to facilitate T-cell proliferation and that the repair of this pathway is also essential in order for LMO2-expressing leukemic clones to undergo abnormal proliferation [142]. Thus, the authors contend that cγ expression is likely required for leukemic progression but is not causative. In support of this conclusion, Scobie et al. have developed a knock-in mouse which expresses cγ under the control of the CD2 promoter and locus control region [164]. Although T-cell development was mildly perturbed in these mice, there was no evidence of increased propensity towards leukemia even when the transgene was crossed onto a tumor prone background. Of note, transduction with an MLV-based vector did not promote leukemogenesis in cγ knock-in cells but was associated with leukemic transformation in $c\gamma^{-/-}$ bone marrow cells which appears to agree with previously described findings using the tumor prone $c\gamma^{-/-}$ model [169].

In addition to SCID-X1, there may be other hematopoietic diseases which are candidates for retroviral gene therapy but which require special consideration with regards to their potential increased sensitivity to insertional mutagenesis. The constitutive overexpression of a transgene which elicits a pro-leukemic effect on processes such as proliferation, differentiation or apoptosis is a likely candidate to promote leukemia in co-operation with retroviral-mediated insertional mutagenesis. As discussed above, this is certainly the case for the retroviral overexpression of HOXB4 [195]. Another example would be the correction of congenital amegakaryocytic anemia resulting from mutations in the thrombopoietin receptor Mpl. Overexpression of Mpl using a gammaretroviral vector incorporating the SFFV promoter/enhancer resulted in chronic myeloproliferative disorder in mice transplanted with transduced cells [188]. This was associated with insertional upregulation of disease-promoting genomic loci such as *Fli1; Sfpi1* and *Klf3*. The use of a weaker cellular promoter or a fraction of the endogenous promoter to drive Mpl expression resulted in genetic correction in the absence of myeloproliferative disorder. Another disease-specific factor that may make cells more susceptible to retroviral-mediated genotoxicity is the association of a disease phenotype with genetic instability. For example, Fanconi anemia (FA) patients suffer from bone marrow failure resulting from inactivating mutations in proteins which propagate an epistatic pathway involved in sensing DNA damage and stress [56, 186]. While the retroviral-mediated delivery of the correcting FA gene has been shown to repair the defective DNA repair pathway, the in vitro manipulation of FA cells which would be required for retroviral transduction has been shown to dramatically reduce the repopulating potential of transduced cells and may result in clastogenic mutations that could promote transformation in collaboration with retroviral-mediated insertional mutagenesis [69, 100, 184]. For instance the evolution of monosomy 7 in the CGD trial is particularly noteworthy given the propensity for FA patients to develop this cytogenetic abnormality as a pre-leukemic clone. Finally, factors relating to the gene therapy protocol may increase the likelihood that transduced cells may succumb to insertional mutagenesis-mediated transformation. One of the longest standing proposed applications for retroviral gene therapy of hematopoietic stem cells is the introduction of a chemotherapeutic drug resistance gene in order to facilitate both the protection of the hematopoietic system in the face of anti-tumor chemotherapy and the selective enrichment of gene modified hematopoietic cells upon drug treatment [114, 189]. While gene modified cells should be protected from the cytotoxic effects of the chemotherapeutic agent, the requirement on them to rapidly expand to replace the ablated non-transduced cells will put a significant proliferative stress on the hematopoietic stem cells and progenitors. In addition, since many of the chemotherapeutic agents used to elicit in vivo selection are DNA damaging agents, it is conceivable that transduced cells may encounter some genotoxic damage resulting from the chemotherapeutic regimen (Reviewed in [115]). These processes may promote the formation of mutations that could act to co-operate with retroviral genotoxicity.

6.6.4 Modification of Vector Architecture to Minimize Insertional Mutagenesis

Based upon the integration site preferences described above, it may be possible to reduce the risk of insertional mutagenesis by using a vector system which is based upon a different parent virus and thus take advantage of properties which are inherent to the wild type virus. However, pre-clinical studies suggest that altering vector architecture may have a greater impact upon the genotoxic potential of a given recombinant vector [122, 124]. Furthermore, these vector modifications could conceivably be applied to any retroviral vector system and so could be combined with a safer integration profile in order to minimize the risk of insertional mutagenesis. There are a number of modifications which can be made to increase safety of retroviral vectors (Fig. 6.2).

The presence of a strong viral enhancer/promoter within the retroviral LTR is a major determinant of the ability of provirus to transactivate a neighboring gene [108, 120]. Thus, the deletion of the viral enhancer/promoter element from the U3 region of the 3′ LTR creates a SIN vector which has been shown to decrease the mutagenic potential of both gammaretroviruses and lentiviruses, even when a strong viral enhancer/promoter such as SFFV is simply shifted from the LTR into an internal configuration [120, 122, 124, 196]. One of the major hurdles to overcome in order to translate this technology to the clinic was the difficulty in producing SIN vectors at a high titer at a clinical scale via transient transfection of producer cells. However, novel vector production techniques including the use of wave bioreactors and optimization of packaging constructs has facilitated the production of SIN vectors at a titer which will facilitate transduction of human $CD34^+$ cells [157, 158].

In order to further reduce the genotoxic potential of SIN vectors, a weaker cellular promoter/enhancer can be incorporated into the internal position of the vector in order to drive expression of the transgene cassette (Fig. 6.2). The use of a weaker internal promoter significantly decreases the ability of both gammaretroviral and lentiviral vectors to transactivate nearby cellular genes [122, 124, 196]. Although cellular promoters may be safer, careful pre-clinical testing is required in order to establish that they are able to drive transgene expression at a level which is therapeutically relevant [117, 176, 194]. Alternatively, the transgene's endogenous promoter could be used to drive its own expression if the regulatory regions have been well enough defined. The minimal regulatory elements of the beta-globin locus control region (LCR) have been successfully cloned into lentiviral vectors in order to mediate high level, erythroid specific expression of globin genes for the treatment of thalassemias and sickle cell anemia [110, 141, 152]. Although the LCR has recently been shown to mediate insertional activation of cellular genes near the insertion site of a recombinant lentiviral vector, this occurred in the context of a mature erythroid cells [71]. It has been hypothesized that the LCR will have little enhancer activity in immature cells which are the likely target for cellular transformation and has subsequently been demonstrated to have low transforming potential compared to internal constitutive promoters [7]. Yet another alternative is the use of an enhancer-less promoter such as the ubiquitous chromatin opening element (UCOE). The UCOE

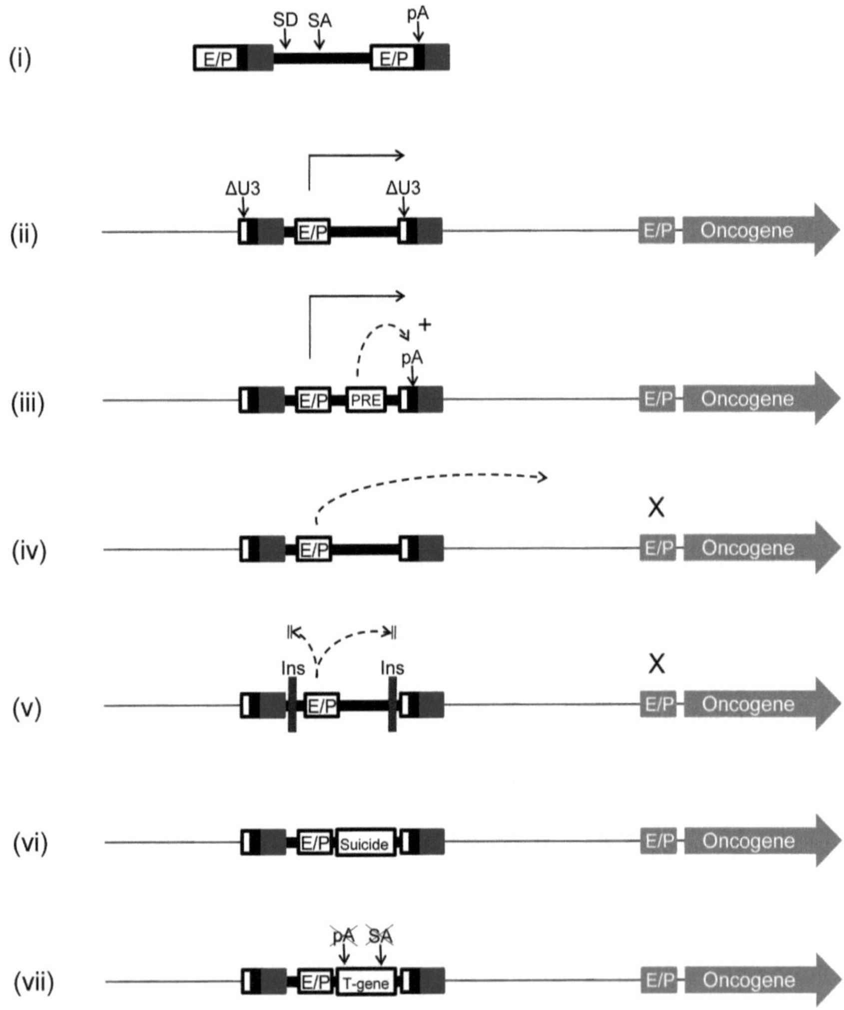

Fig. 6.2 Illustration of the possible modifications that can be made to safety-modify recombinant gene therapy vectors for gene therapy applications. (i) Schematic representation of standard recombinant retroviral vector from Fig. 6.1. (ii) Self-inactivating (SIN) recombinant retroviral vector. The viral enhancer/promoter is deleted from the U3 region of the 3′ LTR resulting in complete absence of this element following reverse transcription and proviral integration. An internal enhancer/promoter drives expression of the transgene cassette. This configuration overcomes the possibility of transcription initiation from the 3′ LTR (see Fig. 6.1b (i)). (iii) SIN vector configuration incorporating an element to enhance polyadenylation in the R region of the 3′ LTR. Enhancing polyadenylation at the 3′ LTR may help overcome insertional mutagenesis via read through transcription (see Fig. 6.1b (ii)) but may increase the potency of vectors to interrupt gene expression if the insertion sites lies within coding exons (see Fig. 6.1b (v)). (iv) SIN vector incorporating either a weaker cellular promoter/enhancer; a lineage specific promoter/enhancer; an inducible promoter/enhancer; or the transgene's endogenous promoter/enhancer. Using a weaker enhancer may eliminate/limit enhancer effect upon genes in the vicinity of the proviral integration site (see Fig. 6.1b (iii & iv)). Using a context specific promoter/enhancer will prevent constitutive trans-activation of genes next to the proviral integration site or will limit their activation to mature

can be incorporated into the internal position of a SIN vector and, even though it is only able to drive very low expression of a transgene, it has been shown to be sufficient to express γc at a level which corrects SCID-X1 [194]. Although the UCOE is unable to transactivate neighboring genes via an enhancer-mediated effect, it remains to be seen whether the maintenance of a permissive chromatin structure has any impact in terms of altering expression of cellular genes. Finally, one could incorporate an inducible promoter into the internal position of a SIN vector. While this system would be highly desirable in terms of allowing the physician to directly regulate the timing and level of transgene expression, inducible systems such as the tetracycline-based system are problematic in terms of achieving tight regulated expression in the context of including all the required regulatory elements within a single retroviral vector [144].

While SIN vectors reduce the risk of enhancer-mediated insertional mutagenesis, they are still susceptible to the potential activation of transcripts downstream of the integration site via leaky polyadenylation in the 3′ LTR R region. There are a number of viral and cellular elements which have been defined to enhance the efficiency of polyadenylation when they are positioned upstream of the polyadenylation signal. Both the woodchuck hepatitis virus post-transcriptional regulatory element (wPRE) and tandem insertions of the SV40 upstream sequence element (USE) have been incorporated into SIN vectors and successfully demonstrated to enhance polyadenylation in the 3′ LTR and minimize read-through transcription [74, 156]. It must be noted that while this strategy may reduce the risk of gene activation via read-through transcription, it has a theoretical risk of increasing the propensity of the vector to interrupt gene expression if the proviral integration site is within a gene's actively transcribed region.

Another commonly proposed strategy to minimize the ability of a retroviral-encoded transcript to transactivate neighboring genes is to incorporate chromatin insulating elements into the recombinant vector flanking the expression cassette.

Fig. 6.2 (continued) cell types which have a decreased transformation potential. (v) Incorporation of insulating elements flanking the promoter/enhancer and transgene expression cassette. The use of insulating elements may prevent trans-activation of neighboring genes via enhancer effect and may additionally make expression of the transgene cassette more robust by forming a barrier between inhibitory chromatin structure and the expression cassette. (vi) Addition of a suicide gene within the co-expression cassette. Co-expression of a pro-drug activating enzyme allows selective killing of gene modified cells upon detection of clonal imbalance/transformation. (vii) Codon optimization of transgene. Conservative mutations can be made in the transgene cDNA in order to remove cryptic splice donor/acceptor sites, enhancer sequences and non-canonical polyadenylation signals such that the provirus has less potential to interfere with endogenous gene expression. Additionally, optimization of codon usage to favor amino acyl-tRNA molecules that are most abundant in mammalian cells will potentiate higher protein expression per recombinant mRNA molecule. ΔU3, LTR U3 region deleted for the sequences encoding the viral enhancer/promoter; E/P, enhancer/promoter; pA, polyadenylation signal; PRE, post-transcriptional regulatory element; +, augmenting effect upon polyadenylation; X, no enhancer effect upon the promoter of endogenous genes; Ins, insulating element; SA, splice acceptor site; T-Suppr, tumor suppressor gene; T-gene, therapeutic transgene

The HS4 chromatin insulator (cHC4) element from the chicken beta globin LCR has been defined as possessing both insulator activity and barrier activity [6, 48, 98]. That is, it can insulate the surrounding genomic locus from the enhancer effect of the retroviral-encoded enhancer while acting as a barrier to the inhibitory effect of the chromatin structure surrounding the insertion site on the retroviral expression cassette. It appears that the full length cHS4 element can compromise vector titer if inserted into the LTR region [7]. Several groups have used either the full length cHS4 element or various truncated versions of this element in order to reduce the transactivating potential of retroviral vectors [7, 49, 99, 151]. Furthermore, the ability of this element to prevent vector silencing and position dependent effects on gene expression may facilitate the incorporation of weaker cellular promoters into SIN vectors while promoting robust expression of the transgene cassette.

If retroviral insertional mutagenesis does result in the generation of a transformed clone, then a mechanism through which the gene modified clone could be selectively eliminated would be highly desirable. Pro-drug activating enzymes such as herpes simplex virus thymidine kinase or bacterial nitoreductase can specifically convert a harmless pro-drug into a toxic metabolite specifically within the cell that the enzyme is expressed and can thus act as a suicide gene. The co-expression of such a suicide within a retroviral gene expression cassette would allow the selective elimination of the gene modified graft upon systemic treatment with the pro-drug [17, 181]. This approach has been somewhat limited by problems relating to adequate expression levels in eukaryotic cells and the immunogenicity of these enzymes. However, site directed mutagenesis to increase enzyme stability/activity has resulted in the derivation of suicide gene which demonstrate increased efficacy [17, 91, 102, 172]. The characterization of mammalian pro-drug activating enzymes such as carboxylesterase may allow the substitution of bacterial/viral enzymes with less immunogenic proteins [109].

Finally, transgene expression cassettes can contain cryptic post-transcriptional regulatory elements such as splice donor/acceptor sites and polyadenylation sites. While these elements will be detrimental in terms of reducing transgene expression level, they also have the potential to mediate insertional mutagenesis via increasing the potential of the vector to interrupt gene expression if the proviral integration site lies in the transcribed region of a gene, as seen in the French β thalassemia trial. Insertional mutagenesis or complete codon optimization can be performed to eliminate these potentially mutagenic sequences [53, 89, 126, 145]. As with the use of insulator elements, this has the added benefit of increasing transgene expression levels and my therefore facilitate the use of weaker promoter elements which have a reduced genotoxic potential.

6.7 Summary

The severe adverse events that have occurred in three gene therapy trials relating to retroviral-mediated insertional mutagenesis have been a setback in the effort to translate the promise of hematopoietic gene therapy into successful clinical therapies. Nonetheless, these unfortunate events have been the catalyst that has driven the field to devise methods to dissect the mechanisms underlying the phenomenon of insertional mutagenesis and to design new strategies to overcome this problem. The resulting technology which has been developed to address this issue does not only have direct application to the field of gene therapy, but may be used to study other biological phenomenon relating to areas such as retroviral pathogenesis, stem cell biology, cellular transformation, and cell signaling.

Acknowledgments We thank Megan Smith, Miriam Edlund and Elise Porter for administrative support and members of our laboratory and the Transatlantic Gene Therapy Consortium for helpful discussions. This work is supported by NIH HL081499.

References

1. Aiuti, A., et al. (2007). Multilineage hematopoietic reconstitution without clonal selection in ADA-SCID patients treated with stem cell gene therapy. *The Journal of Clinical Investigation, 117*, 2233–2240.
2. Aiuti, A., et al. (2002). Correction of ADA-SCID by stem cell gene therapy combined with nonmyeloablative conditioning. *Science, 296*, 2410–2413.
3. Aiuti, A., et al. (2009). Gene therapy for immunodeficiency due to adenosine deaminase deficiency. *The New England Journal of Medicine, 360*, 447–458.
4. Alexander, B. L., et al. (2007). Progress and prospects: Gene therapy clinical trials (part 1). *Gene Therapy, 14*, 1439–1447.
5. Antoine, C., et al. (2003). Long-term survival and transplantation of haemopoietic stem cells for immunodeficiencies: Report of the European experience 1968–1999. *Lancet, 361*, 553–560.
6. Arumugam, P. I., Scholes, J., Perelman, N., Xia, P., Yee, J. K., & Malik, P. (2007). Improved human beta-globin expression from self-inactivating lentiviral vectors carrying the chicken hypersensitive site-4 (cHS4) insulator element. *Molecular Therapy, 15*, 1863–1871.
7. Arumugam, P. I., et al. (2009). Genotoxic potential of lineage-specific lentivirus vectors carrying the beta-globin locus control region. *Molecular Therapy, 17*, 1929–1937.
8. Barjesteh van Waalwijk van Doorn-Khosrovani, S., et al. (2003). High EVI1 expression predicts poor survival in acute myeloid leukemia: A study of 319 de novo AML patients. *Blood, 101*, 837–845.
9. Baum, C., Hegewisch-Becker, S., Eckert, H. G., Stocking, C., & Ostertag, W. (1995). Novel retroviral vectors for efficient expression of the multidrug resistance (MDR-1) gene in early hematopoietic cells. *Journal of Virology, 69*, 7541–7547.
10. Beard, B. C., et al. (2007). Comparison of HIV-derived lentiviral and MLV-based gammaretroviral vector integration sites in primate repopulating cells. *Molecular Therapy, 15*, 1356–1365.
11. Berenson, R. J., et al. (1988). Antigen CD34+ marrow cells engraft lethally irradiated baboons. *The Journal of Clinical Investigation, 81*, 951–955.
12. Berenson, R. J., et al. (1991). Engraftment after infusion of CD34+ marrow cells in patients with breast cancer or neuroblastoma. *Blood, 77*, 1717–1722.
13. Berger, F., et al. (2001). Efficient retrovirus-mediated transduction of primitive human peripheral blood progenitor cells in stroma-free suspension culture. *Gene Therapy, 8*, 687–696.
14. Bertrand, Y., et al. (1999). Influence of severe combined immunodeficiency phenotype on the outcome of HLA non-identical, T-cell-depleted bone marrow transplantation: A retrospective European survey from the European group for bone marrow transplantation and the European society for immunodeficiency. *The Journal of Pediatrics, 134*, 740–748.

15. Beug, H., von Kirchbach, A., Doderlein, G., Conscience, J. F., & Graf, T. (1979). Chicken hematopoietic cells transformed by seven strains of defective avian leukemia viruses display three distinct phenotypes of differentiation. *Cell, 18*, 375–390.
16. Blaese, R. M., et al. (1995). T lymphocyte-directed gene therapy for ADA-SCID: Initial trial results after 4 years. *Science, 270*, 475–480.
17. Blumenthal, M., Skelton, D., Pepper, K. A., Jahn, T., Methangkool, E., & Kohn, D. B. (2007). Effective suicide gene therapy for leukemia in a model of insertional oncogenesis in mice. *Molecular Therapy, 15*, 183–192.
18. Bokhoven, M., et al. (2009). Insertional gene activation by lentiviral and gammaretroviral vectors. *Journal of Virology, 83*, 283–294.
19. Booth, C., et al. (2007). Management options for adenosine deaminase deficiency; proceedings of the EBMT satellite workshop (Hamburg, March 2006. *Clinical Immunology, 123*, 139–147.
20. Bordignon, C., et al. (1995). Gene therapy in peripheral blood lymphocytes and bone marrow for ADA-immunodeficient patients. *Science, 270*, 470–475.
21. Bosticardo, M., Ghosh, A., Du, Y., Jenkins, N., Copeland, N., & Candotti, F. (2009). Self-inactivating retroviral vector-mediated gene transfer induces oncogene activation and immortilization of primary murine bone marrow cells. *Molecular Therapy, 17*, 1910–1918.
22. Bosticardo, M., Marangoni, F., Aiuti, A., Villa, A., & Grazia Roncarolo, M. (2009). Recent advances in understanding the pathophysiology of Wiskott-Aldrich syndrome. *Blood, 113*, 6288–6295.
23. Bousso, P., et al. (2000). Diversity, functionality, and stability of the T cell repertoire derived in vivo from a single human T cell precursor. *Proceedings of the National Academy of Sciences of the USA, 97*, 274–278.
24. Bozorgmehr, F., et al. (2007). No evidence of clonal dominance in primates up to 4 years following transplantation of multidrug resistance 1 retrovirally transduced long-term repopulating cells. *Stem Cells, 25*, 2610–2618.
25. Boztug, K., et al. (2007). Hematopoietic stem cell gene therapy for Wiskott-Aldrich Syndrome. p. 502.
26. Buckley, R. H., et al. (1999). Hematopoietic stem-cell transplantation for the treatment of severe combined immunodeficiency. *The New England Journal of Medicine, 340*, 508–516.
27. Buonamici, S., Chakraborty, S., Senyuk, V., & Nucifora, G. (2003). The role of EVI1 in normal and leukemic cells. *Blood Cells, Molecules and Diseases, 31*, 206–212.
28. Buonamici, S., et al. (2004). EVI1 induces myelodysplastic syndrome in mice. *The Journal of Clinical Investigation, 114*, 713–719.
29. Calmels, B., et al. (2005). Recurrent retroviral vector integration at the Mds1/Evi1 locus in nonhuman primate hematopoietic cells. *Blood, 106*, 2530–2533.
30. Cassani, B., et al. (2009). Integration of retroviral vectors induces minor changes in the transcriptional activity of T cells from ADA-SCID patients treated with gene therapy. *Blood, 114*, 3546–3556.
31. Cattoglio, C., et al. (2007). Hot spots of retroviral integration in human CD34+ hematopoietic cells. *Blood, 110*, 1770–1778.
32. Cavazzana-Calvo, M., & Fischer, A. (2007). Gene therapy for severe combined immunodeficiency: Are we there yet? *The Journal of Clinical Investigation, 117*, 1456–1465.
33. Cavazzana-Calvo, M., Lagresle, C., Hacein-Bey-Abina, S., & Fischer, A. (2005). Gene therapy for severe combined immunodeficiency. *Annual Review of Medicine, 56*, 585–602.
34. Cavazzana-Calvo, M., et al. (2000). Gene therapy of human severe combined immunodeficiency (SCID)-X1 disease [see comments]. *Science, 288*, 669–672.
35. Cavazzana-Calvo, M., et al. (2008). Hematopoietic Stem Cell Gene Therapy Trial with Lentiviral Vector in X-Linked Adrenoleukodystrophy. *American Society of Hematology*, ISI:000262104701045, 304–305.

36. Chong, H., & Vile, R. G. (1996). Replication-competent retrovirus produced by a 'split-function' third generation amphotropic packaging cell line. *Gene Therapy, 3*, 624–629.
37. Cleynen, I., & Van de Ven, W. J. (2008). The HMGA proteins: A myriad of functions [Review]. *International Journal of Oncology, 32*, 289–305.
38. Cornetta, K., & Anderson, W. F. (1989). Protamine sulfate as an effective alternative to polybrene in retroviral-mediated gene-transfer: Implications for human gene therapy. *Journal of Virological Methods, 23*, 187–194.
39. Cornils, K., et al. (2009). Stem cell marking with promotor-deprived self-inactivating retroviral vectors does not lead to induced clonal imbalance. *Molecular Therapy, 17*, 131–143.
40. Couture, L. A., Mullen, C. A., & Morgan, R. A. (1994). Retroviral vectors containing chimeric promoter/enhancer elements exhibit cell-type-specific gene expression. *Human Gene Therapy, 5*, 667–677.
41. Crooks, G. M., & Kohn, D. B. (1993). Growth factors increase amphotropic retrovirus binding to human CD34+ bone marrow progenitor cells. *Blood, 82*, 3290–3297.
42. Daniel, R., & Smith, J. A. (2008). Integration site selection by retroviral vectors: Molecular mechanism and clinical consequences. *Human Gene Therapy, 19*, 557–568.
43. Dave, U. P., et al. (2009). Murine leukemias with retroviral insertions at Lmo2 are predictive of the leukemias induced in SCID-X1 patients following retroviral gene therapy. *PLoS Genetics, 5*, e1000491.
44. De Palma, M., et al. (2005). Promoter trapping reveals significant differences in integration site selection between MLV and HIV vectors in primary hematopoietic cells. *Blood, 105*, 2307–2315.
45. Deichmann, A., et al. (2007). Vector integration is nonrandom and clustered and influences the fate of lymphopoiesis in SCID-X1 gene therapy. *The Journal of Clinical Investigation, 117*, 2225–2232.
46. Dilber, M. S., Bjorkstrand, B., Li, K. J., Smith, C. I., Xanthopoulos, K. G., & Gahrton, G. (1994). Basic fibroblast growth factor increases retroviral-mediated gene transfer into human hematopoietic peripheral blood progenitor cells. *Experimental Hematology, 22*, 1129–1133.
47. Donahue, R. E., et al. (1992). Helper virus induced T cell lymphoma in nonhuman primates after retroviral mediated gene transfer. *The Journal of Experimental Medicine, 176*, 1125.
48. Emery, D. W., Yannaki, E., Tubb, J., Nishino, T., Li, Q., & Stamatoyannopoulos, G. (2002). Development of virus vectors for gene therapy of beta chain hemoglobinopathies: Flanking with a chromatin insulator reduces gamma-globin gene silencing in vivo. *Blood, 100*, 2012–2019.
49. Evans-Galea, M. V., Wielgosz, M. M., Hanawa, H., Srivastava, D. K., & Nienhuis, A. W. (2007). Suppression of clonal dominance in cultured human lymphoid cells by addition of the cHS4 insulator to a lentiviral vector. *Molecular Therapy, 15*, 801–809.
50. Felice, B., et al. (2009). Transcription factor binding sites are genetic determinants of retroviral integration in the human genome. *PloS One, 4*, e4571.
51. Feron, C., et al. (1993). Purification, expansion, and multiple fluorochrome labeling of cord blood hemopoietic precursors: Preliminary results. *Journal of Hematotherapy, 2*, 259–261.
52. Fischer, A., et al. (1997). Naturally occurring primary deficiencies of the immune system. *Annual Review of Immunology, 15*, 93–124.
53. Foster, H., et al. (2008). Codon and mRNA sequence optimization of microdystrophin transgenes improves expression and physiological outcome in dystrophic mdx mice following AAV2/8 gene transfer. *Molecular Therapy, 16*, 1825–1832.
54. Freas-Lutz, D. L., Correll, P. H., Dougherty, S. F., Xu, L., Pluznik, D. H., & Karlsson, S. (1994). Expression of human glucocerebrosidase in murine macrophages: Identification of efficient retroviral vectors. *Experimental Hematology, 22*, 857–865.
55. Garcia, I. S., et al. (1991). A study of chromosome 11p13 translocations involving TCR beta and TCR delta in human T cell leukaemia. *Oncogene, 6*, 577–582.

56. Garcia-Higuera, I., et al. (2001). Interaction of the Fanconi anemia proteins and BRCA1 in a common pathway. *Molecular Cell*, *7*, 249–262.
57. Garrett, E., Miller, A. R., Goldman, J. M., Apperley, J. F., & Melo, J. V. (2000). Characterization of recombination events leading to the production of an ecotropic replication-competent retrovirus in a GP+envAM12-derived producer cell line. *Virology*, *266*, 170–179.
58. Gaspar, H. B., et al. (2004). Gene therapy of X-linked severe combined immunodeficiency by use of a pseudotyped gammaretroviral vector. *Lancet*, *364*, 2181–2187.
59. Gaspar, H. B., et al. (2006). Successful reconstitution of immunity in ADA-SCID by stem cell gene therapy following cessation of PEG-ADA and use of mild preconditioning. *Molecular Therapy*, *14*, 505–513.
60. Glimm, H., et al. (1997). Efficient gene transfer in primitive CD34+/CD38lo human bone marrow cells reselected after long-term exposure to GALV-pseudotyped retroviral vector. *Human Gene Therapy*, *8*, 2079–2086.
61. Goerner, M., Bruno, B., McSweeney, P. A., Buron, G., Storb, R., & Kiem, H. P. (1999). The use of granulocyte colony-stimulating factor during retroviral transduction on fibronectin fragment CH-296 enhances gene transfer into hematopoietic repopulating cells in dogs. *Blood*, *94*, 2287–2292.
62. Gonda, T. J., Sheiness, D. K., & Bishop, J. M. (1982). Transcripts from the cellular homologs of retroviral oncogenes: Distribution among chicken tissues. *Molecular and Cellular Biology*, *2*, 617–624.
63. Grunebaum, E., et al. (2006). Bone marrow transplantation for severe combined immune deficiency. *Jama*, *295*, 508–518.
64. Hacein-Bey-Abina, S., et al. (2002). Sustained correction of X-linked severe combined immunodeficiency by ex vivo gene therapy. *The New England Journal of Medicine*, *346*, 1185–1193.
65. Hacein-Bey-Abina, S., et al. (2003). LMO2-associated clonal T cell proliferation in two patients after gene therapy for SCID-X1. *Science*, *302*, 415–419.
66. Hacein-Bey-Abina, S., et al. (2008). Insertional oncogenesis in 4 patients after retrovirus-mediated gene therapy of SCID-X1. *The Journal of Clinical Investigation*, *118*, 3132–3142.
67. Hacker, C. V., et al. (2006). The integration profile of EIAV-based vectors. *Molecular Therapy*, *14*, 536–545.
68. Haddad, E., et al. (1998). Long-term immune reconstitution and outcome after HLA-nonidentical T-cell-depleted bone marrow transplantation for severe combined immunodeficiency: A European retrospective study of 116 patients. *Blood*, *91*, 3646–3653.
69. Haneline, L. S., et al. (2003). Retroviral-mediated expression of recombinant Fancc enhances the repopulating ability of Fancc–/– hematopoietic stem cells and decreases the risk of clonal evolution. *Blood*, *101*, 1299–1307.
70. Hanenberg, H., Xiao, X. L., Dilloo, D., Hashino, K., Kato, I., & Williams, D. A. (1996). Colocalization of retrovirus and target cells on specific fibronectin fragments increases genetic transduction of mammalian cells. *Nature Medicine*, *2*, 876–882.
71. Hargrove, P. W., et al. (2008). Globin lentiviral vector insertions can perturb the expression of endogenous genes in beta-thalassemic hematopoietic cells. *Molecular Therapy*, *16*, 525–533.
72. Hematti, P., et al. (2004). Distinct genomic integration of MLV and SIV vectors in primate hematopoietic stem and progenitor cells. *PLoS Biology*, *2*, e423.
73. Hershfield, M. S. (1998). Adenosine deaminase deficiency: Clinical expression, molecular basis, and therapy. *Seminars in Hematology*, *35*, 291–298.
74. Higashimoto, T., et al. (2007). The woodchuck hepatitis virus post-transcriptional regulatory element reduces readthrough transcription from retroviral vectors. *Gene Therapy*, *14*, 1298–1304.

75. Hildinger, M., Abel, K. L., Ostertag, W., & Baum, C. (1999). Design of 5′ untranslated sequences in retroviral vectors developed for medical use. *Journal of Virology*, *73*, 4083–4089.
76. Hildinger, M., Eckert, H. G., Schilz, A. J., John, J., Ostertag, W., & Baum, C. (1998). FMEV vectors: Both retroviral long terminal repeat and leader are important for high expression in transduced hematopoietic cells. *Gene Therapy*, *5*, 1575–1579.
77. Hirschhorn, R., Yang, D. R., Puck, J. M., Huie, M. L., Jiang, C. -K., & Kurlandsky, L. E. (1996). Spontaneous in vivo reversion to normal of an inherited mutation in a patient with adenosine deaminase deficiency. *Nature Genetics*, *13*, 290–295.
78. Hock, R. A., Miller, A. D., & Osborne, W. R. (1989). Expression of human adenosine deaminase from various strong promoters after gene transfer into human hematopoietic cell lines. *Blood*, *74*, 876–881.
79. Hoogerbrugge, P. M., et al. (1996). Bone marrow gene transfer in three patients with adenosine deaminase deficiency. *Gene Therapy*, *3*, 179–183.
80. Howe, S. J., et al. (2008). Insertional mutagenesis combined with acquired somatic mutations causes leukemogenesis following gene therapy of SCID-X1 patients. *The Journal of Clinical Investigation*, *118*, 3143–3150.
81. Hu, J., et al. (2003). Direct comparison of RD114-pseudotyped versus amphotropic-pseudotyped retroviral vectors for transduction of rhesus macaque long-term repopulating cells. *Molecular Therapy*, *8*, 611–617.
82. Hu, J., et al. (2008). Reduced genotoxicity of avian sarcoma leukosis virus vectors in rhesus long-term repopulating cells compared to standard murine retrovirus vectors. *Molecular Therapy*, *16*, 1617–1623.
83. Hudspeth, M. P., & Raymond, G. V. (2007). Immunopathogenesis of adrenoleukodystrophy: Current understanding. *Journal of Neuroimmunology*, *182*, 5–12.
84. Kiem, H. P., et al. (1997). Gene transfer into marrow repopulating cells: Comparison between amphotropic and gibbon ape leukemia virus pseudotyped retroviral vectors in a competitive repopulation assay in baboons. *Blood*, *90*, 4638–4645.
85. Kiem, H. -P., et al. (1998). Improved gene transfer into baboon marrow repopulating cells using recombinant human fibronectin fragment CH-296 in combination with IL-6, stem cell factor, FLT-3 ligand, and megakaryocyte growth and development factor. *Blood*, *92*, 1878–1886.
86. Kiem, H. P., et al. (2004). Long-term clinical and molecular follow-up of large animals receiving retrovirally transduced stem and progenitor cells: No progression to clonal hematopoiesis or leukemia. *Molecular Therapy*, *9*, 389–395.
87. Kim, W. S., Weickert, C. S., & Garner, B. (2008). Role of ATP-binding cassette transporters in brain lipid transport and neurological disease. *Journal of Neurochemistry*, *104*, 1145–1166.
88. Kim, Y. J., et al. (2009). Sustained high-level polyclonal hematopoietic marking and transgene expression 4 years after autologous transplantation of rhesus macaques with SIV lentiviral vector-transduced CD34+ cells. *Blood*, *113*, 5434–5443.
89. Knipper, R., et al. (2001). Improved post-transcriptional processing of an MDR1 retrovirus elevates expression of multidrug resistance in primary human hematopoietic cells. *Gene Therapy*, *8*, 239–246.
90. Kohn, D. B., et al. (1995). Engraftment of gene-modified umbilical cord blood cells in neonates with adenosine deaminase deficiency. *Nature Medicine*, *1*, 1017–1023.
91. Kokoris, M. S., Sabo, P., Adman, E. T., & Black, M. E. (1999). Enhancement of tumor ablation by a selected HSV-1 thymidine kinase mutant. *Gene Therapy*, *6*, 1415–1426.
92. Kurre, P., Morris, J., Horn, P. A., Harkey, M. A., Andrews, R. G., & Kiem, H. P. (2001). Gene transfer into baboon repopulating cells: A comparison of Flt-3 Ligand and megakaryocyte growth and development factor versus IL-3 during ex vivo transduction. *Molecular Therapy*, *3*, 920–927.

93. Kustikova, O., et al. (2005). Clonal dominance of hematopoietic stem cells triggered by retroviral gene marking. *Science*, *308*, 1171–1174.
94. Kustikova, O. S., et al. (2007). Retroviral vector insertion sites associated with dominant hematopoietic clones mark "stemness" pathways. *Blood*, *109*, 1897–1907.
95. Laufs, S., Nagy, K. Z., Giordano, F. A., Hotz-Wagenblatt, A., Zeller, W. J., & Fruehauf, S. (2004). Insertion of retroviral vectors in NOD/SCID repopulating human peripheral blood progenitor cells occurs preferentially in the vicinity of transcription start regions and in introns. *Molecular Therapy*, *10*, 874–881.
96. Lewinski, M. K., & Bushman, F. D. (2005). Retroviral DNA integration–mechanism and consequences. *Advances in Genetics*, *55*, 147–181.
97. Lewinski, M. K., et al. (2006). Retroviral DNA integration: Viral and cellular determinants of target-site selection. *PLoS Pathogens*, *2*, e60.
98. Li, C. L., & Emery, D. W. (2008). The cHS4 chromatin insulator reduces gammaretroviral vector silencing by epigenetic modifications of integrated provirus. *Gene Therapy*, *15*, 49–53.
99. Li, C. L., Xiong, D., Stamatoyannopoulos, G., & Emery, D. W. (2009). Genomic and functional assays demonstrate reduced gammaretroviral vector genotoxicity associated with use of the cHS4 chromatin insulator. *Molecular Therapy*, *17*, 716–724.
100. Li, X., et al. (2005). Ex vivo culture of Fancc–/– stem/progenitor cells predisposes cells to undergo apoptosis, and surviving stem/progenitor cells display cytogenetic abnormalities and an increased risk of malignancy. *Blood*, *105*, 3465–3471.
101. Li, Z., et al. (2002). Murine leukemia induced by retroviral gene marking. *Science*, *296*, 497.
102. Likar, Y., et al. (2008). A new acycloguanosine-specific supermutant of herpes simplex virus type 1 thymidine kinase suitable for PET imaging and suicide gene therapy for potential use in patients treated with pyrimidine-based cytotoxic drugs. *Journal of Nuclear Medicine*, *49*, 713–720.
103. Malech, H., et al. (1997). Prolonged production of NADPH oxidase-corrected granulocytes after gene therapy of chronic granulomatous disease. *Proceedings of the National Academy of Sciences of the USA*, *94*, 12133–12138.
104. Malik, P., Krall, W. J., Yu, X. J., Zhou, C., & Kohn, D. B. (1995). Retroviral-mediated gene expression in human myelomonocytic cells: A comparison of hematopoietic cell promoters to viral promoters. *Blood*, *86*, 2993–3005.
105. Mann, R., Mulligan, R. C., & Baltimore, D. (1983). Construction of a retrovirus packaging mutant and its use to produce helper-free defective retrovirus. *Cell*, *33*, 153–159.
106. Manning, J. S., Hackett, A. J., & Darby, N. B., Jr. (1971). Effect of polycations on sensitivity of BALD-3T3 cells to murine leukemia and sarcoma virus infectivity. *Applied Microbiology*, *22*, 1162–1163.
107. Markowitz, D., Goff, S., & Bank, A. (1988). Construction and use of a safe and efficient amphotropic packaging cell line. *J of Virology*, *167*, 400–406.
108. Maruggi, G., et al. (2009). Transcriptional enhancers induce insertional gene deregulation independently from the vector type and design. *Molecular Therapy*, *17*, 851–856.
109. Matzow, T., et al. (2007). Hypoxia-targeted over-expression of carboxylesterase as a means of increasing tumour sensitivity to irinotecan (CPT-11) . *The Journal of Gene Medicine*, *9*, 244–252.
110. May, C., et al. (2000). Therapeutic haemoglobin synthesis in beta-thalassaemic mice expressing lentivirus-encoded human beta-globin. *Nature*, *406*, 82–86.
111. Mazzolari, E., et al. (2007). Long-term immune reconstitution and clinical outcome after stem cell transplantation for severe T-cell immunodeficiency. *Journal of Allergy and Clinical Immunology*, *120*, 892–899.
112. Miller, A. D. (1990). Retrovirus packaging cells. *Human Gene Therapy*, *1*, 5–14.
113. Miller, A. D., Garcia, J. V., von Suhr, N., Lynch, C. M., Wilson, C., & Eiden, M. V. (1991). Construction and properties of retrovirus packaging cells based on gibbon ape leukemia virus. *Journal of Virology*, *65*, 2220–2224.

114. Milsom, M. D., & Fairbairn, L. J. (2004). Protection and selection for gene therapy in the hematopoietic system. *The Journal of Gene Medicine*, *6*, 133–146.
115. Milsom, M. D., & Williams, D. A. (2007). Live and let die: In vivo selection of gene-modified hematopoietic stem cells via MGMT-mediated chemoprotection. *DNA Repair (Amst)*, *6*, 1210–1221.
116. Milsom, M. D., et al. (2005). Overexpression of HOXB4 confers a myelo-erythroid differentiation delay in vitro. *Leukemia*, *19*, 148–153.
117. Milsom, M. D., et al. (2008). Reciprocal relationship between O6-methylguanine-DNA methyltransferase P140K expression level and chemoprotection of hematopoietic stem cells. *Cancer Research*, *68*, 6171–6180.
118. Mitchell, R. S., et al. (2004). Retroviral DNA integration: ASLV, HIV, and MLV show distinct target site preferences. *PLoS Biology*, *2*, E234.
119. Modlich, U., et al. (2005). Leukemias following retroviral transfer of multidrug resistance 1 (MDR1) are driven by combinatorial insertional mutagenesis. *Blood*, *105*, 4235–4246.
120. Modlich, U., et al. (2006). Cell culture assays reveal the importance of retroviral vector design for insertional genotoxicity. *Blood*, *108*, 2545–2553.
121. Modlich, U., et al. (2008). Leukemia induction after a single retroviral vector insertion in Evi1 or Prdm16. *Leukemia*, *22*, 1519–1528.
122. Modlich, U., et al. (2009). Insertional transformation of hematopoietic cells by self-inactivating lentiviral and gammaretroviral vectors. *Molecular Therapy*, *17*, 1919–1928.
123. Montini, E., et al. (2006). Hematopoietic stem cell gene transfer in a tumor-prone mouse model uncovers low genotoxicity of lentiviral vector integration. *Nature Biotechnology*, *24*, 687–696.
124. Montini, E., et al. (2009). The genotoxic potential of retroviral vectors is strongly modulated by vector design and integration site selection in a mouse model of HSC gene therapy. *The Journal of Clinical Investigation*, *119*, 964–975.
125. Moolten, F. L., & Cupples, L. A. (1992). A model for predicting the risk of cancer consequent to retroviral gene therapy. *Human Gene Therapy*, *3*, 479–486.
126. Moreno-Carranza, B., et al. (2009). Transgene optimization significantly improves SIN vector titers, gp91phox expression and reconstitution of superoxide production in X-CGD cells. *Gene Therapy*, *16*, 111–118.
127. Moritz, T., Patel, V. P., & Williams, D. A. (1994). Bone marrow extracellular matrix molecules improve gene transfer into human hematopoietic cells via retroviral vectors. *The Journal of Clinical Investigation*, *93*, 1451–1457.
128. Moser, H. W., Mahmood, A., & Raymond, G. V. (2007). X-linked adrenoleukodystrophy. *Nature Clinical Practice*, *3*, 140–151.
129. Neale, G. A., Rehg, J. E., & Goorha, R. M. (1995). Ectopic expression of rhombotin-2 causes selective expansion of CD4-CD8- lymphocytes in the thymus and T-cell tumors in transgenic mice. *Blood*, *86*, 3060–3071.
130. Neven, B., et al. (2009). Long-term outcome after hematopoietic stem cell transplantation of a single-center cohort of 90 patients with severe combined immunodeficiency. *Blood*, *113*, 4114–4124.
131. Nienhuis, A. (2006). Assays to Evaluate the Genotoxicity of Retroviral Vectors. *Molecular Therapy*, *13*, 461–462.
132. Noguchi, M., et al. (1993). Interleukin-2 receptor gamma chain mutation results in X-linked severe combined immunodeficiency in humans. *Cell*, *73*, 147–157.
133. Nolta, J. A., Crooks, G. M., Overell, R. W., Williams, D. E., & Kohn, D. B. (1992). Retroviral vector-mediated gene transfer into primitive human hematopoietic progenitor cells: Effects of mast cell growth factor (MGF) combined with other cytokines. *Experimental Hematology*, *20*, 1065–1071.
134. Notarangelo, L. D., Miao, C. H., & Ochs, H. D. (2008). Wiskott-Aldrich syndrome. *Current Opinion in Hematology*, *15*, 30–36.

135. Ochs, H. D., Filipovich, A. H., Veys, P., Cowan, M. J., & Kapoor, N. (2008). Wiskott-Aldrich syndrome: Diagnosis, clinical and laboratory manifestations, and treatment. *Biology of Blood and Marrow Transplant, 15*, 84–90.
136. Orkin, S. H., & Nathan, D. G. (1981). The molecular genetics of thalassemia. *Advances in Human Genetics, 11*, 233–280.
137. Ott, M. G., Seger, R., Stein, S., Siler, U., Hoelzer, D., & Grez, M. (2007). Advances in the treatment of chronic granulomatous disease by gene therapy. *Current Gene Therapy, 7*, 155–161.
138. Ott, M. G., et al. (2006). Correction of X-linked chronic granulomatous disease by gene therapy, augmented by insertional activation of MDS1-EVI1, PRDM16 or SETBP1. *Nature Medicine, 12*, 401–409.
139. Ott, M. G., et al. (2007). Phase I/II gene therapy study for Chronic Granulomatous Disease: Results, lessons and perspectives. *American Society of Hematology*, ISI:000251100800504, 503.
140. Otto, E., et al. (1994). Characterization of a replication-competent retrovirus resulting from recombination of packaging and vector sequences. *Human Gene Therapy, 5*, 567–575.
141. Perumbeti, A., et al. (2009). A novel human gamma-globin gene vector for genetic correction of sickle cell anemia in a humanized sickle mouse model: Critical determinants for successful correction. *Blood, 114*, 1174–1185.
142. Pike-Overzet, K., et al. (2006). Gene therapy: Is IL2RG oncogenic in T-cell development? *Nature, 443*, E5, discussion E6–7.
143. Pike-Overzet, K., et al. (2007). Ectopic retroviral expression of LMO2, but not IL2Rgamma, blocks human T-cell development from CD34+ cells: Implications for leukemogenesis in gene therapy. *Leukemia, 21*, 754–763.
144. Pluta, K., Luce, M. J., Bao, L., Agha-Mohammadi, S., & Reiser, J. (2005). Tight control of transgene expression by lentivirus vectors containing second-generation tetracycline-responsive promoters. *The Journal of Gene Medicine, 7*, 803–817.
145. Radcliffe, P. A., et al. (2008). Analysis of factor VIII mediated suppression of lentiviral vector titres. *Gene Therapy, 15*, 289–297.
146. Raper, S. E., et al. (2003). Fatal systemic inflammatory response syndrome in a ornithine transcarbamylase deficient patient following adenoviral gene transfer. *Molecular Genetics and Metabolism, 80*, 148–158.
147. Relander, T., et al. (2005). Gene transfer to repopulating human CD34+ cells using amphotropic-, GALV-, or RD114-pseudotyped HIV-1-based vectors from stable producer cells. *Molecular Therapy, 11*, 452–459.
148. Robinson, D., Elliott, J. F., & Chang, L. J. (1995). Retroviral vector with a CMV-IE/HIV-TAR hybrid LTR gives high basal expression levels and is up-regulated by HIV-1 Tat. *Gene Therapy, 2*, 269–278.
149. Roussel, M., et al. (1979). Three new types of viral oncogene of cellular origin specific for haematopoietic cell transformation. *Nature, 281*, 452–455.
150. Russell, M., et al. (1994). Expression of EVI1 in myelodysplastic syndromes and other hematologic malignancies without 3q26 translocations. *Blood, 84*, 1243–1248.
151. Ryu, B. Y., Evans-Galea, M. V., Gray, J. T., Bodine, D. M., Persons, D. A., & Nienhuis, A. W. (2008). An experimental system for the evaluation of retroviral vector design to diminish the risk for proto-oncogene activation. *Blood, 111*, 1866–1875.
152. Samakoglu, S., et al. (2006). A genetic strategy to treat sickle cell anemia by coregulating globin transgene expression and RNA interference. *Nature Biotechnology, 24*, 89–94.
153. Sandrin, V., et al. (2002). Lentiviral vectors pseudotyped with a modified RD114 envelope glycoprotein show increased stability in sera and augmented transduction of primary lymphocytes and CD34+ cells derived from human and nonhuman primates. *Blood, 100*, 823–832.

154. Sankaran, V. G., et al. (2008). Human fetal hemoglobin expression is regulated by the developmental stage-specific repressor BCL11A. *Science*, *322*, 1839–1842.
155. Sauvageau, G., et al. (1995). Overexpression of HOXB4 in hematopoietic cells causes the selective expansion of more primitive populations in vitro and in vivo. *Genes and Development*, *9*, 1753–1765.
156. Schambach, A., Galla, M., Maetzig, T., Loew, R., & Baum, C. (2007). Improving transcriptional termination of self-inactivating gamma-retroviral and lentiviral vectors. *Molecular Therapy*, *15*, 1167–1173.
157. Schambach, A., Swaney, W. P., & van der Loo, J. C. (2009). Design and production of retro- and lentiviral vectors for gene expression in hematopoietic cells. *Methods in Molecular Biology*, *506*, 191–205.
158. Schambach, A., et al. (2006). Woodchuck hepatitis virus post-transcriptional regulatory element deleted from X protein and promoter sequences enhances retroviral vector titer and expression. *Gene Therapy*, *13*, 641–645.
159. Schiedlmeier, B., et al. (2003). High-level ectopic HOXB4 expression confers a profound in vivo competitive growth advantage on human cord blood CD34+ cells, but impairs lymphomyeloid differentiation. *Blood*, *101*, 1759–1768.
160. Schmidt, M., et al. (2001). Detection and direct genomic sequencing of multiple rare unknown flanking DNA in highly complex samples. *Human Gene Therapy*, *12*, 743–749.
161. Schmidt, M., et al. (2002). Polyclonal long-term repopulating stem cell clones in a primate model. *Blood*, *100*, 2737–2743.
162. Schmidt, M., et al. (2007). High-resolution insertion-site analysis by linear amplification-mediated PCR (LAM-PCR). *Nature Methods*, *4*, 1051–1057.
163. Schwarzwaelder, K., et al. (2007). Gammaretrovirus-mediated correction of SCID-X1 is associated with skewed vector integration site distribution in vivo. *The Journal of Clinical Investigation*, *117*, 2241–2249.
164. Scobie, L., et al. (2009). A novel model of SCID-X1 reconstitution reveals predisposition to retrovirus-induced lymphoma but no evidence of gammaC gene oncogenicity. *Molecular Therapy*, *17*, 1031–1038.
165. Segal, A. W. (1996). The NADPH oxidase and chronic granulomatous disease. *Molecular Medicine Today*, *2*, 129–135.
166. Seger, R. A., et al. (2002). Treatment of chronic granulomatous disease with myeloablative conditioning and an unmodified hemopoietic allograft: A survey of the European experience, 1985–2000. *Blood*, *100*, 4344–4350
167. Seggewiss, R., et al. (2006). Acute myeloid leukemia is associated with retroviral gene transfer to hematopoietic progenitor cells in a rhesus macaque. *Blood*, *107*, 3865–3867.
168. Semmler, A., Kohler, W., Jung, H. H., Weller, M., & Linnebank, M. (2008). Therapy of X-linked adrenoleukodystrophy. *Expert review of Neurotherapeutics*, *8*, 1367–1379.
169. Shou, Y., Ma, Z., Lu, T., & Sorrentino, B. P. (2006). Unique risk factors for insertional mutagenesis in a mouse model of XSCID gene therapy. *Proceedings of the National Academy of Sciences of the USA*, *103*, 11730–11735.
170. Somia, N., & Verma, I. M. (2000). Gene therapy: Trials and tribulations. *Nature Reviews*, *1*, 91–99.
171. Stephan, V., et al. (1996). Atypical X-linked severe combined immunodeficiency due to possible spontaneous reversion of the genetic defect in T cells. *The New England Journal of Medicine*, *335*, 1563–1567.
172. Stolworthy, T. S., et al. (2008). Yeast cytosine deaminase mutants with increased thermostability impart sensitivity to 5-fluorocytosine. *Journal of Molecular Biology*, *377*, 854–869.

173. Suzuki, T., Minehata, K., Akagi, K., Jenkins, N. A., & Copeland, N. G. (2006). Tumor suppressor gene identification using retroviral insertional mutagenesis in Blm-deficient mice. *EMBO Journal*, *25*, 3422–3431.
174. Takeshita, T., et al. (1992). Cloning of the gamma chain of the human IL-2 receptor. *Science*, *257*, 379–382.
175. Tassi, C., Fortuna, A., Bontadini, A., Lemoli, R. M., Gobbi, M., & Tazzari, P. L. (1991). CD34 or S313 positive cells selection by avidin-biotin immunoadsorption. *Haematologica*, *76*(Suppl 1), 41–43.
176. Thornhill, S. I., et al. (2008). Self-inactivating gammaretroviral vectors for gene therapy of X-linked severe combined immunodeficiency. *Molecular Therapy*, *16*, 590–598.
177. Thrasher, A. J., et al. (2006). Gene therapy: X-SCID transgene leukaemogenicity. *Nature*, *443*, E5–6, discussion E6–7.
178. Touw, I. P., & Erkeland, S. J. (2007). Retroviral insertion mutagenesis in mice as a comparative oncogenomics tool to identify disease genes in human leukemia. *Molecular Therapy*, *15*, 13–19.
179. Trasher, A. J., & Gaspar, B. Successful bone marrow gene therapy for SCID. *Blood Cells Molecules, and Diseases*, *40*, 287–288.
180. Trobridge, G. D., et al. (2006). Foamy virus vector integration sites in normal human cells. *Proceedings of the National Academy of Sciences of the USA*, *103*, 1498–1503.
181. Uchiyama, T., et al. (2006). Application of HSVtk suicide gene to X-SCID gene therapy: Ganciclovir treatment offsets gene corrected X-SCID B cells. *Biochemical and Biophysical Research Communications*, *341*, 391–398.
182. Vanin, E. F., Kaloss, M., Broscius, C., & Nienhuis, A. W. (1994). Characterization of replication-competent retroviruses from non-human primates with virus-induced T cell lymphomas and observations regarding the mechanism of oncogenesis. *Journal of Virology*, *68*, 4241–4250.
183. Wagner, W., et al. (2005). Retroviral integration sites correlate with expressed genes in hematopoietic stem cells. *Stem Cells*, *23*, 1050–1058.
184. Walsh, C. E., et al. (1994). A functionally active retrovirus vector for gene therapy in Fanconi anemia group C. *Blood*, *84*, 453–459.
185. Wang, G. P., et al. (2008). DNA bar coding and pyrosequencing to analyze adverse events in therapeutic gene transfer. *Nucleic Acids Research*, *36*, e49.
186. Wang, W. (2007). Emergence of a DNA-damage response network consisting of Fanconi anaemia and BRCA proteins. *Nature Reviews*, *8*, 735–748.
187. Weatherall, D. J. (1998). Pathophysiology of thalassaemia. *Bailliere's Clinical Haematology*, *11*, 127–146.
188. Wicke, D. C., et al. (2008). Ectopic Expression of the Extracellular Domain of Mpl Is Sufficient to Induce a Hematopoietic Population Crisis. *American Society of Hematology*, ISI:000262104703324, 994.
189. Williams, D. A., Lemischka, I. R., Nathan, D. G., & Mulligan, R. C. (1984). Introduction of new genetic material into pluripotent haematopoietic stem cells of the mouse. *Nature*, *310*, 476–480.
190. Winkelstein, J. A., et al. (2000). Chronic granulomatous disease. Report on a national registry of 368 patients. *Medicine*, *79*, 155–169.
191. Woods, N. B., Bottero, V., Schmidt, M., von Kalle, C., & Verma, I. M. (2006). Gene therapy: Therapeutic gene causing lymphoma. *Nature*, *440*, 1123.
192. Wu, X., Li, Y., Crise, B., & Burgess, S. M. (2003). Transcription start regions in the human genome are favored targets for MLV integration. *Science*, *300*, 1749–1751.
193. Yamada, Y., Warren, A. J., Dobson, C., Forster, A., Pannell, R., & Rabbitts, T. H. (1998). The T cell leukemia LIM protein Lmo2 is necessary for adult mouse hematopoiesis. *Proceedings of the National Academy of Sciences of the USA*, *95*, 3890–3895.

194. Zhang, F., et al. (2007). Lentiviral vectors containing an enhancer-less ubiquitously acting chromatin opening element (UCOE) provide highly reproducible and stable transgene expression in hematopoietic cells. *Blood*, *110*, 1448–1457.
195. Zhang, X. B., et al. (2008). High incidence of leukemia in large animals after stem cell gene therapy with a HOXB4-expressing retroviral vector. *The Journal of Clinical Investigation*, *118*, 1502–1510.
196. Zychlinski, D., et al. (2008). Physiological promoters reduce the genotoxic risk of integrating gene vectors. *Molecular Therapy*, *16*, 718–725.

Chapter 7
Bioinformatics of High-Throughput Insertional Mutagenesis

Keiko Akagi, Ming Yi, Jean Roayaei, and Robert M. Stephens

Abstract Bioinformatics plays critical roles to handle large amount of sequence data from insertional mutagenesis. First, computational approaches are used to develop rapid sequence analysis pipelines and biological databases. Millions of reads from an insertion mutagenesis screening are mapped to genomic locations and be annotated to their target genes rapidly by pipeline, and such sequence-based data is stored and managed in database to share the information in the scientific community. Second, statistical techniques are used to distinguish true common insertion sites (loci that have been hit by insertions in multiple tumors: candidate loci for cancer genes) from background insertions in large-scale screenings. Finally, the advanced data mining techniques, pathway and network analysis, are used to give further biological meaning to insertion sites by identifying the interaction of genes in cancer. In this chapter, we discuss features of these three topics and address their future roles: (1) development of sequence analysis pipeline and database, (2) detection of common insertion sites, and (3) network and pathway analysis of insertion sites.

7.1 Sequence Analysis Pipeline and Database

7.1.1 Sequencing and Barcoding Technologies

Insertional mutagenesis is a great tool to find cancer gene candidates from various tumor models [22]. The attempt to screen the insertion sites from retrovirus-induced tumors started by using Southern blotting analysis and genomic library screening in 1980s [31, 70], but the utility of this approach greatly increased after the

K. Akagi (✉)
Mouse Cancer Genetics Program, National Cancer Institute, Center for Cancer Research, Frederick, MD 21702, USA; Comprehensive Cancer Center, Ohio State University, Columbus, OH 43210, USA
e-mail: keiko.akagi@osumc.edu

Sections 7.1 to 7.6 are written by Keiko Akagi and Jean Roayaei and Sections 7.7 to 7.10 are written by Ming Yi and Robert M. Stephens.

A.J. Dupuy, D.A. Largaespada (eds.), *Insertional Mutagenesis Strategies in Cancer Genetics*, DOI 10.1007/978-1-4419-7656-7_7,

development of rapid insert cloning methods and the completion of mouse genome project [43, 47, 66].

Recently, the non-Sanger-based next-generation sequencing technologies enabled us to read the high volume of sequences in a reduced cost [59], and their development will further increase the power of insertional mutagenesis as high throughput screening approach. Currently, three major high-throughput sequence platforms are the Genome Sequencers from Roche/454 Life Sciences [46], the G2 Analyzer from Illumina/Solexa [6], and the SOLiD System from Applied Biosystems (http://solid.appliedbiosystems.com). The comparison between three platforms show a trade-offs between average sequence read length and the number of sequence reads produced in a single run. The Illumina and SOLiD systems generate ten-fold more reads than Roche/454 Sequencers, but they provide much shorter read length of 35–50 bp than that of Roche/454 Sequencers, 250 bp [57, 59]. The longer read length of Roche/454 Sequencers makes the system ideal for insertional mutagenesis screenings with a barcoding approach. Historically, "barcodes", unique DNA sequence identifiers, have been used in EST-based gene discovery projects to distinguish tissue sources of pooled cDNA libraries [26]. In insertional mutagenesis screenings, researchers analyze multiple tumors and/or tissues from many individuals, and the number of necessary insertion reads in each sample is much smaller than the capacity of the Roche/454 sequencer. To overcome the limited number of samples sequenced in parallel by the physical separators of sequencer, barcodes (~10 bp) are assigned to each insertional mutagenesis sample. The presence of barcodes allows independent samples to be pooled together for sequencing and allows the bioinformatic segregation of the output from sequencers afterwards. The length of the barcodes should be determined to allow the unambiguous assignment of barcodes and to minimize the base reads used for barcodes. For Roche/454 sequencers, the design of barcodes should consider the pyrosequencing error characteristics. For example, to avoid homopolymeric runs in pyrosequencing technologies, the barcodes should avoid including more than two successive occurrences of the same nucleotides for 10 bp barcodes [52]. With the careful designed barcodes, ~98% of the sequences can be assigned to a correct sample [38]. This approach greatly enhances the saturation of insertional mutagenesis screenings.

7.2 Sequence Mapping

Collected sequences from insertional mutagenesis are mapped to the genome assembly of model organisms to identify the genomic characteristics of insertion sites. The mouse genome was sequenced from C57BL/6 J mouse strain, and the assembled genome was distributed to three genome browsers (UCSC, NCBI, and Ensembl) [74]. The genome sequence assembly in these genome browsers is updated periodically (the latest release: NCBI build m37, October 2007), and genomic coordinates along the chromosome can change dramatically in different assemblies because of changes in sequence data or algorithm implementation. When there is a doubt of error in a region of the assembly, it is worth to review the mapping record of insertion sites for older versions of the genome assembly. NCBI also provides limited genome assemblies from alternative mouse strains (mixed strains by Celera whole genome shot-gun sequencing and 129 substrains) [50], and Sanger Institute is attempting de nove assembly of sequences of 17 mouse strain genomes on Illumina's GA platform [63]. These alternative assemblies can be used to map insertional mutagenesis sequences to strain specific regions.

To align sequences of insertion sites to genome, two sequence alignment programs are used frequently: BLAST (Basic Local Alignment Search Tool) [3] and BLAT (BLAST-like Alignment Tool) [39]. Both programs find regions of local similarity between sequences, and they rapidly scans for short sequence matches and extends these matches into high scoring pairs. However, the BLAT algorithm builds an index of the genome assemblies before any searches are performed to achieve faster speed than BLAST does. The BLAT program is sometimes not as sensitive as BLAST, but it is much faster and often used to find the position of an insertion site sequence in genome.

Once mapped, the insertion site sequence is checked for the alignment at the junction of mutagens (such as retrovirus' long terminal repeat) to identify the position and direction of insertion in genome. The direction of insertion site helps to identify the target genes because mutagens affect genes more severely in certain gene locations and directions. For example, retrovirus insertions are more likely to enhance expression of genes when they are located in the upstream of genes in the antisense orientation or downstream in the sense orientation [72]. Mutagenic Sleeping Beauty transposon vectors are designed to affect genes in a directional manner [14, 23]. In addition to the direction of insertion site, the number of redundant sequences, how many redundant sequences arose from the same tumor and mapped to the same position in the genome, also provides useful information. In a shotgun-cloning project, insertions with higher redundancy are more likely to be clonal and early insertions in tumorigenesis by Southern blot [19]. If there is no severe bias in cloning step, the frequency of redundant read for an insertion site serve as an indicator of the copy number of insertion, penetrance, in the tumor library.

7.3 Target Gene Identification

After the mapping of insertion sequences to genome, the annotation of these insertions sites to neighboring target genes becomes an important step. RefSeq is a crated non-redundant collection of genomes, transcripts, and proteins [54]. The genes in RefSeq are supported by experimental evidence and multiple quality assessment tests; the gene set is ideal to annotate insertion sites to confirmed genes. Recently, reports show the involvement of microRNAs and non-coding RNAs in retrovirus insertion sites [17, 34, 61], and some microRNAs are suggested to act as oncogenes in human cancer [8]. The miRBase [28], database for published microRNAs, is another useful gene set to identify the potential involvement of microRNAs in insertion sites.

In addition to gene databases, medical and phenotype databases contribute to the process of target gene identification for insertion sites. To identify whether the target gene of insertion site is a known human cancer gene, Cancer Gene Census works as a great resource. The Cancer Genome Project at Sanger Institute conducted a census of cancer-causing genes from literature [25], and they provide the list of cancer genes and their mutation/tumor types (367 gene; October 2007 release). More than 40 of these loci are also identified as common insertion sites from the retrovirus insertional mutagenesis screening [72]. To search non-cancerous phenotypes in human, Online Mendelian Inheritance in Man (OMIM) [30] provides information about human genes and their disease phenotypes. To identify the phenotype in mice, Jackson Laboratory's Mouse Genome Database [11] provides information about phenotypes for gene-target mouse.

7.4 Insertional Mutagenesis Database

To efficiently store and manage sequences and annotated data, a relational database plays an important role for insertional mutagenesis. A relational database is a database that organizes data into tables and represents relationships among those tables, and these relationships enable users to combine data from multiple tables [18]. To manage relational databases, there are many database management systems available. Popular systems are Oracle (http://www.oracle.com), MySQL (http://www.mysql.com), and PostgreSQL (http://www.postgresql.org). In typical insertional mutagenesis screenings, a database requires following tables to perform basic functions: (1) genotype or phenotype information of mice, (2) histology/pathology report of tumors, (3) raw and processed sequence of insertion sites, (4) mapping information of insertion sites, and (5) annotation information on mapped insertion sites. Each table is linked to one or more table through key columns such as mouse ID, tumor ID, and sequence ID, and SQL queries are performed to merge distinct tumor data sets and to facilitate annotation process.

To share the data of insertional mutagenesis in wide scientific community, several groups created web-accessible databases [1, 56, 65]. One such database, Retroviral Tagged Cancer Gene Database (http://rtcgd.ncifcrf.gov), was created to provide the opportunity to analyze multiple data sets from mouse tumor models to be analyzed as one [1]. This database contains a collection of more than 15 retroviral insertion screens and 2 transposon insertion screens for cancer genes. Currently, there are approximately 14 000 insertions derived from 1900 tumors in the database, and the tally is regularly updated to include new screening sets from recent publications. This resource provides tools to compare different screenings from different mouse backgrounds and to find interactions between cancer genes. In addition, these datasets have been linked to other biological resources and been annotated onto the mouse genome browsers at UCSC (http://genome.ucsc.edu) to reflect the latest genome annotation around the insertion sites.

There are a couple of ideas to enhance the insertional mutagenesis resources. First, current resources show only cancer gene candidates identified in each screening set. It would be useful to introduce a tool to reveal the cancer gene candidates across the multiple screening sets. Second, the resources lack a tool to identify cancer gene candidates those are highly specific to a specific tumor type or tumor model. For example, it is useful to find genes significantly enriched in B-cell tumors compared to T-cell tumors. Similarly, as an extension of this tool, a tool to identify specific biological pathways in specific tumor type and model would be useful, too. Finally, the current and future resources should collect more information about tumors from data source. Adding high-throughput data from various microarray technologies, such as gene expression and copy number variation, to database will further enhance the understanding of effect of insertions in tumorigenesis.

7.5 Detection of Common Insertion Sites

7.5.1 Statistical Approaches of Common Insertion Sites Detection

Common insertion sites (CISs) are loci that have been hit by insertions in multiple tumors. Since insertions occur independently in each tumor, it is unlikely to find the same locus from multiple tumors by chance alone. Such insertion hotspots are more likely to be the result of selective advantage of tumor cells by these insertion loci, and CISs are highly correlated to cancer-causing disease genes [22]. In the early days, researchers screened CISs from small number of insertions and confirmed them by southern blot method. However, when researchers started to perform large-scale insertional mutagenesis screenings by PCR-based methods, the definition of CISs started to raise statistical questions. Since the total size of mouse/human genome stays constant, it is more likely a screening contains random/background CISs when the screening collects large number of insertions. In addition, since the probability of observing common insertion sites depends on the number of possible pairings of genomic slots, the probability increases very rapidly when the size of screening grows [24]. To analyze large-scale screenings for CISs, it became necessary to attach statistical significance to the definition of CISs, and three statistical approaches are developed to control the probability of finding false CISs.

First, Mikkers et al. used mathematical approach to define CISs [47]. They assumed the random insertions in genome follow the Poisson distribution and defined CISs as two insertions within 26 kb or three insertions within 236 kb for a set of 500 retroviral insertions. The basic statistical assumption here is to treat random common insertions in the mouse genome follow a Poisson probability distributions [7]. These insertions are occurring at a constant rate say, at per unit of time. They divide each unit of time into N subunits and generate a random number to determine whether a new CIS has arrived. They statistically generate these times using an exponential probability distribution. The approach is effective to calculate the expected numbers of CISs based on formula quickly, but this method does not allow taking into account of preferential insertion loci.

Next, Suzuki et al. performed computer simulations to assess non-randomness of insertion sites in tumors [66]. They created 1200 random insertions into a genome (2.6×10^6 kb in length) and counted the number of insertions in various window sizes for 100 000 Monte Carlo trials. Based on these computer simulations, they used 30, 50, or 100 kb window sized for two insertions, three insertions, or four insertions, respectively. Monte Carlo Simulation is a statistical method of iteratively evaluating a deterministic model (a model that does not include a random error term) using sets of random numbers as inputs or explanatory variables [75]. This technique is used when the deterministic model is highly nonlinear or includes several parameters subject to variation. The Monte Carlo technique is one way to analyze uncertainties introduced into a deterministic model. The statistical aim is to determine how random variations, lack of a priori knowledge, or random error affects the sensitivity, precision or reliability of the biological system that is being statistically modeled. This approach allows precise modeling of genomic insertions, but the computation takes long time to identify the ideal window sizes for different screening sets.

Finally, Jeron de Ridder et al. have used a statistical technique called "Kernel Density Estimation" to locate the regions in the genome that show a significant increase in insertion density [21]. In previous two methods, one, two, or three predetermined window sizes are used to detect CISs. In contrast, they have implemented this technique at different scales to be able to evaluate the data set at any relevant scale. For any position in the genome, an estimate of the number of insertions can be calculated by summing all the kernel functions. Applying a kernel convolution would smooth the observed insertion in a region close to this insertion, and using kernel convolution removes the problems associated with the data sparseness in genomic data sets. The authors have used a Gaussian probability distribution as kernel function to locate CISs. Their approach is an improvement over ad hoc and non-axiomatic probability models such as using different simulation models that are not reproducible and repeatable for other biological scientists to follow. The authors also attempted the background correction for murine leukemia virus (MLV) insertions based on their preference near transcription start sites.

7.6 Bias of Insertion Site Preference in Genome

To distinguish cancer-causing CISs from background CISs, researchers need to know the bias of insertion site preferences of mutagens. Both retroviruses and transposons have distinctive preferred targeting regions in genome [12]. As for Sleeping Beauty (SB) transposon, it targets TA motif widely in genome and shows less insertion site bias toward genes than retroviruses do [73], [78] #18. SB transposon is an attractive mutagen for disease gene screens, but it has a strong preference to transpose around the donor concatamer (local hopping). Because of this nature, the SB insertion sites mapped to the same chromosome as the donor concatamer cannot be distinguished from background CISs easily. As for retroviruses, the situation is more serious. For example, murine leukemia virus (MLV) so strongly favors promoters of active genes [76] that CIS by chance alone can happen more frequently than expected by the statistical consideration [77]. To detect CISs from highly non-random insertions, the biological characteristics of insertions and CISs should be utilized.

First, it is necessary to build more realistic assumption of background insertions than random insertions in entire genome. Ideally, researchers should sample the large numbers of unselected insertion sites from the mutagen treated host-cell type for their screenings. When it is difficult to obtain the experimental data for unselected insertions, the estimated distribution should be build from the known preference of mutagen. For example, since MLV insertions tend to occur near promoter regions of active genes, the distribution of MLV insertions can be estimated based on genomic information such as known transcription start sites, expression levels of transcripts, and chromatin structure in host-cells. Based on the realistic background distribution of insertions, statisticians can calculate the number of CISs by chance and can adjust the significance estimates of CISs for hotspots.

Second, the frequency of insertion reads (number of redundant reads from high-throughput sequencing) can work as a filter for CIS detection. Current statistical approaches treat all insertion sites equally, but the frequency of insertion reads in tumor can be used to weight or prioritize insertion sites in statistics. A cancer causing insertion happens early in the tumor development and has many copies in tumor library. Although there is a chance that a non-oncogenic insertion is piggybacked to a cancer-causing insertion in tumor, insertions with many redundant reads are more likely to play important roles in tumorigenesis than insertions represented by only one read. By using the frequency of insertion reads, it is possible to filter or eliminate non-oncogenic insertions from high-throughput sequencing screens.

Finally, the patterns of insertion site directions can be an indicator for selection of cancerous mutation. In retrovirus tagged cancer gene database (RTCGD), common insertion sites (CISs) show strong orientation bias relative to genes. For example, 70% of retrovirus CISs are antisense oriented in $5'$ of genes while 45% of non-CIS retrovirus insertions are antisense oriented in $5'$ of genes (Table 7.1). As for SB transposon insertions in RTCGD, 75% (75/99) of CISs are sense oriented in inside of genes while 51% (102/199) of insertions in control embryo samples are sense oriented in inside of gene. Since retroviruses and transposons show no orientation bias relative to genes at the time of insertion (i.e. 50% in sense direction) [5], these direction biases are likely the selection result of tumorigenesis. When an insertion cluster shows patterned insertion orientation (such as a cluster of five insertions with 100% antisense orientation at $5'$ location), the cluster has higher potential to be a true cancer-causing CIS. By applying simple statistics (such as binomial statistics), the significance of orientation bias in a CIS cluster is evaluated and can be used to predict the biological importance of the CIS.

Table 7.1. Location and orientation of retrovirus insertion sites relative to RefSeq

A. Retrovirus insertions not marked as CISs in RTCGD

	5 prime		inside		3 prime	
direction	Frequency	Total (%)	Frequency	Total (%)	Frequency	Total (%)
sense	747	54.61	909	51.62	344	46.30
antisense	621	45.39	852	48.38	399	53.70
Total	1368	100.00	1761	100.00	743	100.00

B. Retrovirus insertions marked as CISs in RTCGD

	5 prime		inside		3 prime	
direction	Frequency	Total (%)	Frequency	Total (%)	Frequency	Total (%)
sense	314	29.65	416	55.03	298	62.08
antisense	745	70.35	340	44.97	182	37.92
Total	1059	100.00	756	100.00	480	100.00

7.7 Pathway and Network Analysis

Although retrovirus-based high throughput (HTP) insertional mutagenesis and genetic screening technologies have received a great deal of attention in past few years, many of the studies have been have primarily focused on the large numbers of identified common insertion sites (CISs) individually. In addition, those insertions that were found in only a single tumor are generally considered as random events and not considered further in terms of their potential biological impacts on tumorigenesis. In recent years, pathway and biological network analysis have emerged as categories of promising analysis methods for HTP data. Such an analysis method is getting more and more attention in genomics and other "omics" fields in both academic and industrial settings. The advances in algorithms, methodology and databases in pathway analysis provide a great opportunity for retrieval and interpretation of underlying biological themes from the insertional mutagenesis screening data in an integrative way. This holds even greater promise because of the possibility of considering genes proximal to the insertion sites not only at the individual gene basis as in conventional CISs, but also at the level of pathways, gene sets, and even biological networks. Furthermore, it is possible that the rare insertion sites not previously considered can be included together for possible additional insights. This chapter will focus specifically on pathway and network analysis of insertional mutagenesis screening data. First, we will provide a brief overview of concepts and methodologies of pathway and network analysis. Then, we will describe the potential applications of pathway and network analysis to insertional mutagenesis screening data. Lastly, we will describe the perspectives of integrative discovery by combining microarray data (expression, CGH) and other HTP technologies with insertional mutagenesis screening data for pathway and network analysis.

7.8 Overview of Pathway and Network Analysis

Pathways have been conventionally used for display, representation and interpretation of biological processes in a visual fashion in biomedical research for a long time. At the molecular level, a picture of an active cell or organism can be envisioned as an interconnected network of molecular components, which consists of functional units in topologies characteristic of pathways functionally linked to one another as biological processes of certain well-defined cellular functions. Everyone with coursework in biochemistry will remember these diagrams. This networked view of biology, along with pathway-level details brings the biological context for systematic understanding of molecular systems of living entities.

Since the earliest pathway visualization web interface of Kyoto Encyclopedia of Genes and Genomes (KEGG) [36, 49], many well-defined pathways were curated into a variety of pathway databases including the BioCarta pathway collections (www.biocarta.com/genes/allpathways.asp) and the Science STKE Cell Signaling pathway database (http://stke.sciencemag.org/cm). These databases are often hand-curated by the actual communities performing various research projects so there is considerable validation associated with the data. Until recently, software tools were implemented to incorporate pathway content mainly for the purpose of visualizing and analyzing high-throughput data including but not limited to GenMAPP [16], Pathway Processor [29], Pathway Tool [37], Cytoscape [60], ViMac [44], Osprey [10], WPS [79], PathSys [4]. Among them, Cytoscape has gained increased popularity largely due to the contributions from many groups and the application still continues to evolve.

Most of the pathway analysis tools focus more on visualization of data in the context of pathways and/or networks, and only some tools may provide statistical assessment of the reliability of each differentially expressed gene [29]. Among them, WPS was probably the first tool that allowed simultaneous visualization of multiple HTP data in the context of one or multiple pathways [79]. WPS is also probably the first tool that attempted to integrate analysis results from enrichment analysis or over-representation analysis (ORA) with networks of genes and associated pathways or terms.

More recently, part of the collection of pathway and network analysis tools also came from commercial sources including Pathway Studio [51]; a product of Ariadne Genomics, www.ariadnegenomics.com), PathArt (a product of Jubilant Biosys Ltd, www.jubilantbiosys.com), Ingenuity Pathways Analysis tool (a product of Ingenuity Systems Inc, www.ingenuity.com), MetaCore (a product of GeneGO Inc, www.genego.com) and the Genomatix software suite GmbH (a product of Genomatix, www.genomatix.de). These tools provide a variety of interfaces for the visualization of gene networks, natural language processing (NLP) extracted, or hand-curated biological pathway/association network databases from literature mining, and usually accepted gene-list based data input for data integration. Some of the tools have nice integration with enrichment analysis and promoter analysis such as the Genomatix promoter analysis module MatInspector [13] and BiblioSphere module [58] as well as the Ingenuity pathway tool.

A conventional analysis scheme using these tools usually begins with identifying differentiated genes using statistical methods and tools [80]. The next step is typically to use pathway/network-based tools to map these identified genes into the context of pathways or biological networks and seek the connection of these genes within pathways and networks for clues of embedded biological themes by means of color cues for data integration, gene-gene or gene-term association relations implicated by connections of nodes in pathways or networks, and higher levels of network features such as hubs derived from the graphical layout of network architecture. Some tools have the capacity to use the results of enrichment analysis to filter and simplify the networks to focus on specific sub-domains of the networks or sub-networks including WPS [79]. Others such as Cytoscape [60], MintViewer [81], and Osprey [10] that focus more on network views and queries of the data, have included features for viewing and querying larger subsets of the networks of association relations such as the interactome on a more global scale. One tool named VisANT [33], which attempts to integrate interactome data from different sources, has the ability to uncover orthologous networks, and perform exploratory data mining and basic graph operations on arbitrary networks and sub-networks, including loop detection, degree distribution (the distribution of edges per node) and shortest path identification between various component genes or proteins. Some very recent efforts have focused more on exploring the topology and architectures of the networks in conjunction with high-throughput data to seek biological scenarios [42, 71].

It should be pointed out that due to the relatively large number of network and pathway tools available for biologists and bioinformatitians to use, that making the decision of which should be used for each application can be a complex process that often involves personal preferences as much as anything else. Side-by-side comparisons among these tools should be made to help users make the best use of the tools. Some comparative efforts have been made [64] [79]. Such a comparison task is complicated by the fact that many of these applications are considered "works in progress" with additional features and capabilities being added almost daily.

As a close relative to pathway/network based analysis approaches, there are a category of related but different methods based upon pre-declared gene-sets. A gene set is a collection of genes that have some functional relevance or relationship (either known or hypothetical) that are put together as a group or a set, which were annotated as representing a certain biological meaning (e.g., GO terms; GSEA annotation terms). In the broader pathway definition, a pathway consists of not only a set of genes, but also includes some physical connection or gene-to-gene relations, usually presented graphically. In contrast to a pathway-based method, a gene-set based approach only considers the fact that the genes in a defined gene set are grouped together based on their association with an annotation term, common functional or structural feature, but without regard to information of any direct relationship such as protein-protein interaction, kinase versus substrate etc.

Gene-set based analysis methods, also referred to as modular methods, include but are not restricted to enrichment-based analysis (or over-representation analysis, ORA), functional class scoring (FCS), global tests, and a singular value decomposition or SVD-based method.

Over-representation analysis (ORA), also called enrichment analysis is a very popular current approach that begins with pre-defined gene lists (e.g. differential genes between tumor and normal tissues), which are subjected to analysis for enrichment levels that evaluate which functional categories are represented in the lists more than expected by chance. These methods are usually based on a one-tailed Fisher's exact test [2, 32, 79]. Many software tools have implemented this algorithm including but not limited to EASE/DAVID [20, 32], GOminer [82]; Fatigo [2], T-profiler [9], WPS [79]. One caveat for such methods is that since they present a ranked list of terms based on the Fisher's exact test p-values or enrichment type scores, they are quite sensitive to the cutoff value used for getting the gene lists [53] initially.

As an alternative, functional class scoring (FCS) usually starts with all genes from a dataset that are then ranked based on their expression differences in terms of statistical significance (e.g., *t*-test *p*-values between the two classes and *p*-value based aggregate scoring FCS method [53]), based on their expression differences in terms of fold change versus a normal distribution [40], or based on correlation levels between their expression in the two classes (e.g., SNR, the signal to noise ratio in GSEA method) [47, 62, 67], or based on correlated expression pattern [41]. Once these rankings, statistical values, or correlation levels of individual genes are derived, aggregate class scores for functional categories are derived with different algorithms that rank the functional terms. The most popular one among this category is probably the GSEA method (www.broad.mit.edu/gsea). The GSEA method has recently been improved by the addition of more GSEA annotation terms as well as the ability to deal with terms or gene sets of different sizes [48, 62], since it was argued by others that the GSEA method may be biased toward assigning higher enrichment scores to gene sets of large size [17]. Such methods have demonstrated increased sensitivity in detecting subtle consistent changes within a gene-set across samples of a class relative to another class that were missed by methods that consider only single gene-based enrichment. This suggests that further gains the ability to draw insights may be made by additional approaches that view individual gene changes in a pathway context more generally (see below).

Other group testing methods include the global test method, which looks for associations between the global expression pattern for a group of genes and a variable of interest (e.g., a clinical outcome) [27]. A singular value decomposition or SVD-based method has been developed that uses the first metagene derived from singular value decomposition (SVD) as the basis for calculation of a defined pathway activity level [69].

7.9 Pathway and Network Analysis of Insertional Mutagenesis Screening Data

Early work on insertional mutagenesis screening has been mostly focused on common insertional sites (CISs), which have led to identification of many genes that are involved in mouse tumorigenesis. Such genes are well documented in the RTCGD database (http://RTCGD.ncifcrf.gov) [1]. Recently, high-throughput screening has significantly increased the amount of data and at the same time also increased the challenge and the burden of data analysis. Ultimately, the goal of such HTP screening is to not only identify as many CISs as possible to uncover the proximal genes of these CISs individually, but it may be even more important to put all the pieces together to interpret the underlying biology and the connections among these genes at CISs and hopefully generate insights for tumorigenesis mechanisms. Pathway and network analysis provides a great opportunity to fulfill such a goal. Evidence has shown that many pathway/gene set based analysis methods or group testing methods identify the same pathways that had already been shown to be involved in the pathogenesis of prostate cancer derived from different prostate datasets, and these pathways/gene sets appeared to be more consistent than simple gene signatures [45]. More interestingly, with the help of pathway and network analysis, it is even possible to include the insertions found in only a single tumor that have previously been ignored in the analysis.

In conventional analysis methods for high throughput data such as that produced by microarrays, gene signatures or differentiated genes can be generated based on their consistent behavior across the sample population. Gene signatures have usually been referred to as a set of genes whose change in behavior (e.g., transcription level, protein expression levels) reflects the change of biological states or stages of disease progression. Some of these gene signatures have been used as biomarkers or therapeutic target lists. Usually, signature genes were identified and then confirmed in follow-up studies. In other cases, signature genes have been widely used to develop classifiers or predictive models for the purpose of diagnostic class prediction. Similar to the conventional way of identifying gene signatures or differentiated genes, the genes at or near CISs were selected as significant genes from insertional mutagenesis screening. The screening procedure is based on the repeated and consistent incidences that the virus insertion sites occur around these genes in independent tumors significantly more than expected by chance [35, 47, 66]. These genes can be treated as "signature" genes, which in many cases can be biologically relevant to the tumorigenesis that is caused by these viral insertions and their corresponding impact on the proximal genes. However, although efforts have been made to use computational approaches to correct for the increased probability of finding false CISs as the amount of available data increases [21], it is still possible that some of these CISs may solely reflect the "easy" or "open accessible" sites for viral insertions in the genome, which may not be the real "causes" for the underlying tumorigenesis mechanisms we are searching for. These CISs have been classified as bystander integrations and result from the fact that many of the tumors analyzed contain multiple integration sites so that the real culprit is not always immediately apparent. In contrast, some other CISs or even rare single insertion sites may have a real impact on tumorigenesis-relevant genes by working together at the level of a pathway or gene set in each individual. These rare sites in concert could in some cases be the real biological processes relevant to tumorigenesis that we are searching for. In other words, it may be the common biological processes, pathways, or gene sets that lead to the tumorigenesis, which are not necessarily common at the gene level or at the level of insertion sites.

Cancer is a complex disease thought to be caused by many biological and genetic events that break through the multiple checkpoints required for tumor progression. Genes that are involved in these multiple checkpoints or biological processes are not necessarily interfered with by the viral CISs, considering the fact that although these insertional mutagenesis screenings were performed at high throughput scale in the sense of over the whole genome, but the number of subjects or mice used in the screenings were still largely limited and relatively small compared to a large number of genes and insertional sites in the genome. This fact would cause many of the genes residing nearby CISs and involved in tumorigenesis, which would be observed as CISs in a larger population, to be observed only as rare or single insertion sites in current relatively small sample population due to the scope of study and/or other factors such as budget constrains. Thus, some of the CISs may represent false discovery due to the existence of "easy" or "open accessible" viral insertion sites in the genome, whereas some rare insertion sites would be biologically relevant, which were "downgraded" from "CISs" due to sample size. CISs-based approach alone may be limited by the "random" and biologically unrelated but "easy" insertion sites in the genome and consequently may not be useful for uncovering the whole repertoire of tumorigenesis mechanisms. Fortunately, pathway and network analysis would be able to help put all these pieces together and uncover the underlying tumorigenesis mechanisms, not only at level of individual genes initially relying on the CISs, but also more commonly than the CISs at the level of pathways or gene sets by taking account of these known CISs as well as genes that are influenced by other rare insertion sites.

Indeed, there are some great advances that have been made recently with pathway-level efforts. In one of the studies, CIS-associated genes were first identified and associated pathways such as Ras signaling pathways and Notch pathways were uncovered [68]. Then genes that were never linked to cancer were also further pursued and some of them such as Rspo3 turned out to strongly enhance oncogenicity [68], suggesting it is always a possibility that genes and pathways that were not related to cancer previously may be indeed involved in oncogensis. The enrichment levels of common gene families and protein domains among the genes associated with RISs (Retroviral integration sites) were analyzed and many of gene families that are known to be involved in cancer were identified as the result of such enrichment analysis. In addition, the pathways and cellular processes that these RIS-associated genes are known to be involved in could also be verified through commercially and publicly accessible pathway and/or network analysis tools such as the Ingenuity Pathway Analysis tool mentioned earlier [68]. In addition, tagged genes in three different tumors are compared. Microarray data and qRT-PCR for expression analysis were performed to confirm the de-regulation of these CISs or RISs associated genes in general in mouse tumors [68]. Although several similar studies exist, this study is a very nice example of how pathway and network analysis can have significant impact on interpreting the screening results at both the individual gene and pathway levels. Since pathway and network analysis has a much wider conceptual range and many categories of functional annotations that consider the combined functionality of multiple genes as a functional group, biological themes that could have been missed from gene-level or CISs-based analysis now can be retrieved as novel underlying biological themes.

Despite the recent effort using pathway analysis in the above example, there are more comprehensive ways to consider the same screening data. First of all, it would be more helpful to have more thorough comparison of different tagged genes in three different tumors at the pathway-level, not just in a limited table format. Secondly, for the genes that are associated with CISs and even the genes associated with the single insertion site, although this study did use enrichment analysis to identify enriched gene families and signaling pathways, the combined gene list was used for the analysis. This would give only an overall estimation of the involved genes of the insertion sites in the sample population of the entire data set screened using the enrichment levels of signaling pathways. However, it should be noted that each of the tumor samples obviously had derived the ability to produce a tumor and thus must have altered sufficient genes through its insertion sites for the transformation process to occur in the first place. In addition, one insertion at the proximal region of one or two critical genes in a pathway could change the entire pathway. Since such screening only looks over much lower numbers of genes at the viral insertion sites, it is very important for the analysis to take account of both ListHits (number of genes from the interested gene list hit or are annotated in a pathway) and enrichment level (with a significantly higher than expected number of genes annotated in the pathway from the interested gene list, in this case, genes coming from the screening) consistency across the screening sample population for the intended pathway would be a better choice compared to simple enrichment analysis on the overall gene list from the insertion sites. All of these analyses can be potentially done in one single software tool: a Pathway Pattern Extraction pipeline (Yi and Stephens, unpublished work on Pathway Pattern Extraction pipeline). This Pathway Pattern Extraction pipeline can collect the ListHits, enrichment levels, or even FDRs of each individual sample screening for a pathways and then combined into a matrix file for pattern extraction at the pathway-level and look for the commonly hit pathways with ListHits, or common enriched pathways with enrichment scores (or levels) or FDRs. This pipeline is flexible in setting up the template for extraction of patterns at pathway level across lists using ListHits or enrichment scores. For example, as for the screening data mentioned above, the comparison of different tagged genes in three different tumors at the pathway-level can be used with this pipeline. Common or unique "hit" or enriched pathways can be extracted and used to create heatmap type views for the ListHits or enrichment level across the different tumors and this is much easier to visualize than tabular format produced with other methods. Using this method, the details of each pathway either common to all or unique to individual tumors can be examined more closely for their associated genes in each tumor. In addition, sample-wise gene lists that were associated with CISs and even the genes associated with the single insertion sites in each screening sample can be analyzed in a similar way with this pipeline, considering for their ListHits and enrichment level of pathways and combined for consistency analysis or pattern extraction at pathway-level. Such an analysis scheme in combination with what has been done in the study [68] could set up an alternative way to analyze insertional mutagenesis data in general.

With the help of large pathway and gene-gene association annotation knowledge databases, either from literature mining or from hand-curation efforts in many commercial and academic pathway and network analysis tools mentioned earlier, such an analysis scheme can be done in a more integrative fashion. As seen in the study mentioned as above [68] it is common to use the whole gene lists derived from CISs and even single insertion sites to generate gene-gene association network for pathway and network analysis, and then analyze the network as a whole for embedded biological themes from the gene-gene associations and functional category enrichments. That analysis gives us the underlying biology from the entire screening experiment as a whole. However, each individual screening sample developed a tumor on their own on top of the viral insertion sites within the genome of corresponding sample. Between these individual tumors, they may share common genes from CISs, but even more likely they may share the same biological processes or pathways that the insertions on each individual have impacts. One way to capture this additional information is to use the Pathway Pattern Extraction Pipeline applied directly to each insertion site-associated gene list from each screening tumor sample as described earlier (Yi and Stephens, unpublished work on Pathway Pattern Extraction pipeline). An alternative way would be to generate association networks from each insertion site-associated gene list from each screening tumor sample, and try to compare the embedded biology combined within these networks. In spite of a lack of solid algorithms and methods for direct comparison of association networks derived from different studies or corresponding gene lists, the comparison can be done based on enrichment level for pathways or gene sets similarly using the Pathway Pattern Extraction pipeline [82]. However, with extra genes derived from the association networks, which were built with the original insertion site-associated gene list, based on curated association relations, such comparisons can be further performed using the associated genes derived from search of the annotation enriched knowledge databases, which extends the scope of genes just at the proximal regions of the insertion sites of screening to their associated genes at large from other studies in the field based on literature. This effort would connect the local gene contexts from screening with the global gene contexts that were derived from a much large scale analysis in the literature. This analysis scheme is expected to help deepen the pathway or network context from the local domain to the much wider global domain and would help uncover novel pathways that previously were not known to be involved in tumorigenesis. We have performed a preliminary analysis based on this method and which demonstrated its effectiveness [19]

7.10 Integrative Discovery with Other Sources of HTP Data

Insertional mutagenesis screening usually only retrieves a limited number of genes at the proximal regions of CISs. Although one insertion at the proximal region of a critical gene of a pathway or biological process could have the potential to change the entire relevant pathway, many of the changes that occur in a pathway would not be revealed without the help of other high-throughput technology such as microarray or proteomics. Evidence has shown that with the help of other sources of data such as genome-wide expression data measured by microarray technology or high throughput qRT-PCR, or even clinicopathological data for the purpose of validation and verification, the discovery would be largely consolidated with much greater confidence [68]. Although public data (e.g., Microarray data previously published or stored in Gene Expression Omnibus or GEO database: http://www.ncbi.nlm.nih.gov/geo) could be an easily accessible data resource used for integrative discovery [68], it would be better if it is feasible to perform microarray experiments on the same tumor subjects that come from the screening. The microarray experiment on the screening subjects would provide direct evidence on the interference of genes by proximal viral insertion based on the change of their expression levels, provided they were compared with expression data derived from the same mouse strain. In addition, since microarray measures expression levels at the genome-wide scale, the responses of other related genes such as genes in the same pathway or under control of the interfered gene if it is a transcription factor could have been also measured in the same microarray data for side-by-side comparison. Such data would give first-hand evidence that a viral insertion would generate a cascade of responses at transcription level that are either at the scope of a pathway or as co-regulated by a transcription factor that was impacted by the viral insertion. If compatible microarray data in terms of tissue type or tumor type is available from public data repository such as GEO database, they can be downloaded and used for analysis along with the screening data in a similar way as described et al [68]. Other resources include Oncomine database [55] and ArrayExpress However, since the microarray data and screening are not derived from the same subject and each tumor more likely would behave differently at individual gene level, only overall impacts can be assessed at more of population level in between the downloaded microarray data and screening data. Only the microarray data from the same screening subject with each tumor would match with the specific viral insertion scheme in the corresponding tumor and would measure the specific response from such viral insertion scheme in this tumor. One of the potential issues in such a situation is that microarray data usually has to compare something as baseline expression such as tumor vs. normal, treated vs. untreated. This may be overcome by using the tumor and normal tissue of the same type from the same screening mouse. In addition, due to the high throughput feature of microarray technology, it is usually performed with either technical or biological replicates. In our screening experiment, it would be hard to have "true" biological replicates since most of the tumors may have different schemes of insertion sites in large portion of viral insertion sites in spite of existence of CISs. However, they can be treated as biological replicates as the same class in comparison with the normal tissues as baseline class, such that in this case, the common transcription responses among the tumor population could be uncovered in line with the genes of CISs. Alternatively, each tumor would be treated differently, but with the help of technical replicates it is possible to uncover the specific transcription response for each individual tumor with a specific viral insertion scheme. The combination of each of the viral insertion positions in individual tumors would provide clues to the complete repertoire of tumorigenesis mechanisms. Therefore, depending on how to use the microarray data, different aspects of the study would be addressed accordingly. Furthermore, although one viral insertion as one of the CISs at the proximal region of a critical gene of a tumorigenesis-related pathway or biological process could be one common initiating process for tumorgenesis in general, at the individual level of tumor samples, the changes of proximal genes caused by multiple rare insertion sites could have synergistic effect at a pathway level when multiple genes at the pathway were targeted by the corresponding viral insertions. Genome-wide transcription level survey by expression microarray could largely address and confirm such incidences along with the powerful pathway and network analysis. An additional integrative approach that would help to tease apart the various components contributing to the initiation, maintenance and progression of tumors would be to study the same impact on mouse expression patterns resulting from controlled expression of know oncogenes in the same context. For example, if similar impacts were observed in expression changes between one of the CISs and expression of v-myc, then any expression effects not seen in the v-myc mouse might suggest additional processes being altered through possibly an additional CIS in that mouse.

The microarray data mentioned above primarily referred to expression array results. Advances in microarray technology including other array types such as CGH (comparative genomic hybridization) array, microRNA array, and ChIP-chip (chromatin immunoprecipitation on chip) array or ChIP-Seq (chromatin immuno-precipitation with the Next Generation Sequencing technology) would provide great opportunity to dissect in depth the CISs and their impacts on genomic stability, miRNA expression, promoter of critical genes involved in tumorgenesis, as well as involved pathways at different levels of regulatory mechanisms. In particular, most recently, the promising Next Generation Sequencing technology for sequencing of genome, transcriptome and epigenome would provide higher resolution and greater coverage of the insertional mutagenesis screening than the array counterparts with decreasing cost.

Although an obvious impact that the viral insertion sites would have is to change the expression pattern of the neighborhood gene(s) directly, there might be other levels of regulation. For example, if a viral insertion, whether it is one of the CISs or rare insertion site, would change the regulation of a proximal miRNA cluster in the genome, it would in turn change the patterns of potential gene targets of the corresponding miRNA(s). This scenario can be carefully investigated with help of miRNA array. Alternatively, if a viral insertion influenced a significant change on the expression of a proximal transcription factor, say p53, with help of ChIP-chip array or Chip-Seq for p53 protein, the target genes of this transcription factors could be refined and validated in genome-wide scale, which may help understand the specific target genes that are essentially for tumorigenesis. In addition, in-silicon promoter analysis of potential target genes of a transcription factor that is targeted by a proximal viral insertion could also provide insight on what impacts the viral insertion could have on biological processes at both individual gene and pathway levels. As mentioned earlier, Genomatix promoter analysis module MatInspector [13] in combination with BiblioSphere [58] would be a good choice to do such promoter analysis. With the help of CGH array or Next Generation Sequencing of genome, the CNV (copy number variation) can be studied tumor by tumor, which would be particularly useful to study the possibility whether viral insertions could cause genome instability, e.g., loss or gain of particular genome region that could reside potential oncogenic or tumor suppressor genes. Such hypothetical CNVs triggered by viral insertions could be a critical tumorgenesis mechanism as well. Other HTP technologies developed and improved rapidly in recent years such as proteomics can provide more useful information from different aspects of gene regulation as well as signaling events at pathway or network contexts.

7.11 Conclusions

Over the course of their relatively short history, pathway analysis tools have evolved into a set of powerful methods for placing data into the context of their associated biology. More recently, the field of pathway and network analysis has achieved a dramatic enhancement in capabilities through conceptual expansion of both the methodologies they employ, but also the expansion of the sources of biological databases and increasly available pathway information that they utilize to gain biological insights. Pathway and network analysis now represent a major way to analyze high throughput data through pathway signatures, pathway-level patterns, pathway-level consistencies, which have been proven to be increasingly effective in uncovering biological themes. Pathway signatures and pathway classifiers derived from datasets for a particular study may turn out to be more generic than gene signatures and classifiers and may be more useful when used to understand underlying biological mechanisms such as tumorigenesis mechanisms. In addition, biological changes at different levels of regulation, including transcription, protein expression, and phosphorylation occurring in the same individual samples, which now can be measured with all types of HTP technologies as they become increasly feasible, could be more likely integrated for discovery of underlying biological themes under the context of pathways, gene sets, biological modules, as well as complicated networks. Therefore, the integrative discovery approach that employs pathway and network analysis as its base of methodology and biology knowledge enriched pathway databases as a base of contexts in combination of other sources of HTP data derived from various microarray technologies, and other HTP platforms including the promising next-generation sequencing technologies, would be the ultimate way to elevate the approach of viral insertional mutagenesis screening up to the systems biology level in so-called post-genomic era. Clearly, one of the advantages of using mouse as an animal model for cancer is the ability to design experiments that leverage both the inbred nature of the mouse giving them a much more uniform genetic background than has been observed in the human population, but also the available reagents such as retroviral insertional mutagenesis and associated controlled oncogene expression studies to uncover clues about the many processes that must be altered in cells through the initiation, maintenance and progression of cancer.

References

1. Akagi, K., Suzuki, T., Stephens, R. M., Jenkins, N. A., & Copeland, N. G. (2004). RTCGD: Retroviral tagged cancer gene database. *Nucleic Acids Research*, *32*(Database issue), D523–D527.
2. Al-Shahrour, F., Díaz-Uriarte, R., & Dopazo, J. (2004). FatiGO: A web tool for finding significant associations of Gene Ontology terms with groups of genes. *Bioinformatics*, *20*, 578–580.
3. Altschul, S. F., Gish, W., et al. (1990). Basic local alignment search tool. *Journal of Molecular Biology*, *215*(3), 403–410.
4. Baitaluk, M., Qian, X., Godbole, S., Raval, A., Ray, A., & Gupta, A. (2006). PathSys: Integrating molecular interaction graphs for systems biology. *BMC Bioinformatics.*, *7*, 55.
5. Barr, S. D., Leipzig, J., et al. (2005). Integration targeting by avian sarcoma-leukosis virus and human immunodeficiency virus in the chicken genome. *Journal of Virology*, *79*(18), 12035–12044.
6. Bennett, S. T., Barnes, C., et al. (2005). Toward the 1,000 dollars human genome. *Pharmacogenomics*, *6*(4), 373–382.
7. Berger, R., & Casella, G. (2001, June 18). *Statistical inferences* (2nd ed., p. 92). Pacific Grove, CA: Duxbury Press.
8. Blenkiron, C., & Miska, E. A. (2007). miRNAs in cancer: Approaches, aetiology, diagnostics and therapy. *Human Molecular Genetics*, *16*(Spec No 1), R106–R113.
9. Boorsma, A., Foat, B. C., Vis, D., Klis, F., & Bussemaker, H. J. (2005). T-profiler: Scoring the activity of predefined groups of genes using gene expression data. *Nucleic Acids Research*, *33*(web server issue), W592–W595.
10. Breitkreutz, B. J., Stark, C., & Tyers, M. (2003). Osprey: A network visualization system. *Genome Biology*, *4*(3), R22.
11. Bult, C. J., Eppig, J. T., et al. (2008). The mouse genome database (MGD): Mouse biology and model systems. *Nucleic Acids Research*, *36*(Database issue), D724–D728.
12. Bushman, F. D. (2003). Targeting survival: Integration site selection by retroviruses and LTR-retrotransposons. *Cell*, *115*(2), 135–138.
13. Cartharius, K., Frech, K., Grote, K., Klocke, B., Haltmeier, M., Klingenhoff, A., et al. (2005). MatInspector and beyond: Promoter analysis based on transcription factor binding sites. *Bioinformatics*, *21*(13), 2933–2942.
14. Collier, L. S., Carlson, C. M., et al. (2005). Cancer gene discovery in solid tumours using transposon-based somatic mutagenesis in the mouse. *Nature*, *436*(7048), 272–276.
15. Cui, J. W., Li, Y. J., et al. (2007). Retroviral insertional activation of the Fli-3 locus in erythroleukemias encoding a cluster of microRNAs that convert Epo-induced differentiation to proliferation. *Blood*, *110*(7), 2631–2640.
16. Dahlquist, K. D., Salomonis, N., Vranizan, K., Lawlor, S. C., & Conklin, B. R. (2002). GenMAPP, a new tool for viewing and analyzing microarray data on biological pathways. *Nature Genetics*, *31*(1), 19–20.
17. Damian, D., & Gorfine, M. (2004). Statistical concerns about the GSEA procedure. *Nature Genetics*, *36*(7), 663.
18. Date, J. C. (2003). Introduction to Database Systems.
19. Dave, U. P., Akagi, K., et al. (2009). Murine leukemias with retroviral insertions at Lmo2 are predictive of the leukemias induced in SCID-X1 patients following retroviral gene therapy. *PLoS Genetics*, *5*(5), e1000491.
20. Dennis, G., Jr., Sherman, B. T., Hosack, D. A., Yang, J., Gao, W., Lane, H. C., et al. (2003). DAVID: Database for annotation, visualization, and integrated discovery. *Genome Biology*, *4*(5), P3.
21. de Ridder, J., Uren, A., Kool, J., Reinders, M., & Wessels, L. (2006). Detecting statistically significant common insertion sites in retroviral insertional mutagenesis screens. *PLoS Computational Biology*, *2*(12), e166.

22. Dudley, J. P. (2003). Tag, you're hit: Retroviral insertions identify genes involved in cancer. *Trends in Molecular Medicine*, *9*(2), 43–45.
23. Dupuy, A. J., Akagi, K., Largaespada, D. A., Copeland, N. G., & Jenkins, N. A. (2005, July 14). Mammalian mutagenesis using a highly mobile somatic Sleeping Beauty transposon system. *Nature, 436*(7048), 221–226. PubMed PMID: 16015321.
24. Fehse, B., & Roeder, I. (2008). Insertional mutagenesis and clonal dominance: Biological and statistical considerations. *Gene Therapy*, *15*(2), 143–153.
25. Futreal, P. A., Coin, L., et al. (2004). A census of human cancer genes. *Nature Reviews Cancer*, *4*(3), 177–183.
26. Gavin, A. J., Scheetz, T. E., et al. (2002). Pooled library tissue tags for EST-based gene discovery. *Bioinformatics*, *18*(9), 1162–1166.
27. Goeman, J. J., van de Geer, S. A., de Kort, F., & van Houwelingen, H. C. (2004). A global test for groups of genes: Testing association with a clinical outcome. *Bioinformatics*, *20*, 93–99.
28. Griffiths-Jones, S. (2006). miRBase: The microRNA sequence database. *Methods in Molecular Biology*, *342*, 129–138.
29. Grosu, P., Townsend, J. P., Hartl, D. L., & Cavalieri, D. (2002). Pathway processor: A tool for integrating whole-genome expression results into metabolic networks. *Genome Research*, *12*(7), 1121–1126.
30. Hamosh, A., Scott, A. F., et al. (2005). Online Mendelian inheritance in man (OMIM), a knowledgebase of human genes and genetic disorders. *Nucleic Acids Research*, *33*(Database issue), D514–D517.
31. Hayward, W. S., Neel, B. G., et al. (1981). Activation of a cellular onc gene by promoter insertion in ALV-induced lymphoid leukosis. *Nature*, *290*(5806), 475–480.
32. Hosack, D. A., Dennis, G., Jr., Sherman, B. T., Lane, H. C., & Lempicki, R. A. (2003). Identifying biological themes within lists of genes with EASE. *Genome Biology*, *4*, R70.
33. Hu, Z., Mellor, J., Wu, J., & DeLisi, C. (2004). VisANT: An online visualization and analysis tool for biological interaction data. *BMC Bioinformatics*, *5*, 17.
34. Huppi, K., Volfovsky, N., et al. (2007). MicroRNAs and genomic instability. *Seminars in Cancer Biology*, *17*(1), 65–73.
35. Johansson, F. K., et al. (2004). Identification of candidate cancer-causing genes in mouse brain tumor by retroviral tagging. *Proceedings of the National Academy of Sciences of the USA*, *101*, 11334–11337.
36. Kanehisa, M., Goto, S., Kawashima, S., & Nakaya, A. (2002). The KEGG databases at GenomeNet. *Nucleic Acids Research*, *30*, 42–46.
37. Karp, P. D., Paley, S., & Romero, P. (2002). The pathway tools software. *Bioinformatics*, *18*, S225–S232.
38. Keng, V. W., Villanueva, A., Chiang, D. Y., Dupuy, A. J., Ryan, B. J., Matise, I., et al. (2009, March). A conditional transposon-basedinsertional mutagenesis screen for genes associated with mouse hepatocellular carcinoma. *Nature Biotechnology, 27*(3), 264–274. Epub 2009 February 22. PubMed PMID: 19234449; PubMed Central PMCID: PMC2712727.
39. Kent, W. J. (2002). BLAT—the BLAST-like alignment tool. *Genome Research*, *12*(4), 656–664.
40. Kim, S. -Y., & Volsky, D. (2005). PAGE: Parametric analysis of gene set enrichment. *BMC Bioinformatics*, *6*, 144.
41. Lamb, L., Ramaswamy, S., Ford, H. L., Contreras, B., Martinez, R. V., Kittrell, F. S., et al. (2003). A mechanism of cyclin D1 action encoded in the patterns of gene expression in human cancer. *Cell*, *114*, 323–334.
42. Lu, X., Jain, V. V., Finn, P. W., & Perkins, D. L. (2007). Hubs in biological interaction networks exhibit low changes in expression in experimental asthma. *Molecular Systems Biology*, *3*, 98.
43. Lund, A. H., Turner, G., et al. (2002). Genome-wide retroviral insertional tagging of genes involved in cancer in Cdkn2a-deficient mice. *Nature Genetics*, *32*(1), 160–165.
44. Luyf, A. C. M., de Cast, J., & van Kampen, A. H. C. (2002). Visualizing metabolic activity on a genome-wide scale. *Bioinformatics*, *18*, 813–818.

45. Manoli, T., Gretz, N., Grone, H. -J., Kenzelmann, M., Elis, R., & Brors, B. (2006). Group testing for pathway analysis improves comparability of different microarray datasets. *Bioinformatics*, *22*, 2500–2506.
46. Margulies, M., Egholm, M., et al. (2005). Genome sequencing in microfabricated high-density picolitre reactors. *Nature*, *437*(7057), 376–380.
47. Mikkers, H., Allen, J., et al. (2002). High-throughput retroviral tagging to identify components of specific signaling pathways in cancer. *Nature Genetics*, *32*(1), 153–159.
48. Mootha, V. K., Lindgren, C. M., Eriksson, K. F., Subramanian, A., Sihag, S., Lehar, J., et al. (2003). PGC-1-responsive genes involved in oxidative phosphorylation are coordinately downregulated in human diabetes. *Nature Genetics*, *34*, 267–273.
49. Nakao, M., Bono, H., Kawashima, S., Kamiya, T., Sato, K., Goto, S., et al. (1999). Genome-scale gene expression analysis and pathway reconstruction in KEGG. *Genome Inform Ser Workshop*, *10*, 94–103.
50. NCBI Mouse Genome Resources. (2008). http://www.ncbi.nlm.nih.gov/genome/guide/mouse, NCBI Bulid 37(2008/08).
51. Nikitin, A., Egorov, S., Daraselia, N., & Mazo, I. (2003). Pathway studio—the analysis and navigation of molecular networks. *Bioinformatics*, *19*(16), 2155–2157.
52. Parameswaran, P., Jalili, R., et al. (2007). A pyrosequencing-tailored nucleotide barcode design unveils opportunities for large-scale sample multiplexing. *Nucleic Acids Research*, *35*(19), e130.
53. Pavlidis, P., Qin, J., Arango, V., Mann, J. J., & Sibille, E. (2004). Using the gene ontology for microarray data mining: A comparison of methods and application to age effects in human prefrontal cortex. *Neurochemical Research*, *29*, 1213–1222.
54. Pruitt, K. D., Tatusova, T., et al. (2007). NCBI reference sequences (RefSeq): A curated non-redundant sequence database of genomes, transcripts and proteins. *Nucleic Acids Research*, *35*(Database issue), D61–D65.
55. Rhodes, D. R., Yu, J., Shanker, K., Deshpande, N., Varambally, R., Ghosh, D., et al. (2004). ONCOMINE: A Cancer Microarray Database and Data-Mining Platform. *Neoplasia*, *6*(1), 1–6.
56. Roberg-Perez, K., Carlson, C. M., et al. (2003). MTID: A database of Sleeping Beauty transposon insertions in mice. *Nucleic Acids Research*, *31*(1), 78–81.
57. Shendure, J., & Ji, H. (2008). Next-generation DNA sequencing. *Nature Biotechnology*, *26*(10), 1135–1145.
58. Scherf, M., Epple, A., & Werner, T. (2005). The next generation of literature analysis: Integration of genomic analysis into text mining. *Briefings in Bioinformatics*, *6*(3), 287–297.
59. Schuster, S. C. (2008). Next-generation sequencing transforms today's biology. *Nature Methods*, *5*(1), 16–18.
60. Shannon, P., Markiel, A., Ozier, O., Baliga, N. S., Wang, J. T., Ramage, D., et al. (2003). Cytoscape: A software environment for integrated models of biomolecular interaction networks. *Genome Research*, *13*, 2498–2504.
61. Slape, C., Hartung, H., et al. (2007). Retroviral insertional mutagenesis identifies genes that collaborate with NUP98-HOXD13 during leukemic transformation. *Cancer Research*, *67*(11), 5148–5155.
62. Subramanian, A., Tamayo, P., Mootha, V., Mukherjee, S., Ebert, B. L., Gillette, M. A., et al. (2005). Gene set enrichment analysis: A knowledge-based approach for interpreting genome-wide expression profiles. *Proceedings of the National Academy of Sciences of the USA*, *102*, 15545–15550.
63. Sudbery, I., Stalker, J., et al. (2009). Deep short-read sequencing of chromosome 17 from the mouse strains A/J and CAST/Ei identifies significant germline variation and candidate genes that regulate liver triglyceride levels. *Genome Biology*, *10*(10), R112.
64. Suderman, M., & Hallett, M. (2007, October 15). Tools for visually exploring biological networks. *Bioinformatics*, *23*(20), 2651–2659. Epub 2007 August 25. Review. PubMed PMID:17720984.

65. Sun, L. V., Jin, K., et al. (2008). PBmice: An integrated database system of piggyBac (PB) insertional mutations and their characterizations in mice. *Nucleic Acids Research*, *36*(Database issue), D729–D734.
66. Suzuki, T., Shen, H., et al. (2002). New genes involved in cancer identified by retroviral tagging. *Nature Genetics*, *32*(1), 166–174.
67. Sweet-Cordero, A., Mukherjee, S., Subramanian, A., You, H., Roix, J. J., Ladd-Acosta, C., et al. (2005). An oncogenic KRAS2 expression signature identified by cross-species gene-expression analysis. *Nature Genetics*, *37*, 48–55.
68. Theodorou, V., Kimm, M. A., Boer, M., Wessels, L., Theelen, W., Jonkers, J., et al. (2007). MMTV insertional mutagenesis identifies genes, gene families and pathways involved in mammary cancer. *Nature Genetics*, *39*(6), 759–769.
69. Tomfohr, J., Lu, J., & Kepler, T. B. (2005). Pathway level analysis of gene expression using singular value decomposition. *BMC Bioinformatics*, *6*, 225.
70. Tsichlis, P. N., Strauss, P. G., et al. (1983). A common region for proviral DNA integration in MoMuLV-induced rat thymic lymphomas. *Nature*, *302*(5907), 445–449.
71. Ulitsky, I., & Shamir, R. (2007). Identification of functional modules using network topology and high-throughput data. *BMC Systems Biology*, *1*, 8.
72. Uren, A. G., Kool, J., et al. (2005). Retroviral insertional mutagenesis: Past, present and future. *Oncogene*, *24*(52), 7656–7672.
73. Vigdal, T. J., Kaufman, C. D., et al. (2002). Common physical properties of DNA affecting target site selection of sleeping beauty and other Tc1/mariner transposable elements. *Journal of Molecular Biology*, *323*(3), 441–452.
74. Waterston, R. H., Lindblad-Toh, K., et al. (2002). Initial sequencing and comparative analysis of the mouse genome. *Nature*, *420*(6915), 520–562.
75. Weisstein, E. W. (2004). Monte carlo method. Retrieved October 06, 2008, from http://mathworld.wolfram.com/MonteCarloMethod.html.
76. Wu, X., Li, Y., et al. (2003). Transcription start regions in the human genome are favored targets for MLV integration. *Science*, *300*(5626), 1749–1751.
77. Wu, X., Luke, B. T., et al. (2006). Redefining the common insertion site. *Virology*, *344*(2), 292–295.
78. Yant, S. R., Wu, X., Huang, Y., Garrison, B., Burgess, S. M., & Kay, M. A. (2005, March). High-resolution genome-wide mapping of transposon integration in mammals. *Molecular and Cellular Biology*, *25*(6), 2085–2094. PubMed PMID: 15743807; PubMed Central PMCID: PMC1061620.
79. Yi, M., Horton, J. D., Cohen, J. C., Hobbs, H. H., & Stephens, R. M. (2006). WholePathwayScope: A comprehensive pathway-based analysis tool for high-throughput data. *BMC Bioinformatics*, *7*, 30.
80. Yi, M., Mudunuri, U., Che, A., & Stephens, R. M. (2009, June 29). Seeking unique and common biological themes in multiple gene lists or datasets: Pathway pattern extraction pipeline for pathway-level comparative analysis. *BMC Bioinformatics, 10*, 200. PubMed PMID: 19563622; PubMed Central PMCID: PMC2709625.
81. Zanzoni, A., Montecchi-Palazzi, L., Quondam, M., Ausiello, G., Helmer-Citterich, M., & Cesareni, G. (2002). MINT: A molecular interaction database. *FEBS Letters*, *513*, 135–140.
82. Zeeberg, B. R., Feng, W., Wang, G., Wang, M. D., Fojo, A. T., Sunshine, M., et al. (2003). GoMiner: A resource for biological interpretation of genomic and proteomic data. *Genome Biology*, *4*, R28.

Index

Note: The letters 'f' and 't' following the locators refer to figures and tables respectively.

A.J. Dupuy, D.A. Largaespada (eds.), *Insertional Mutagenesis Strategies in Cancer Genetics*, DOI 10.1007/978-1-4419-7656-7,

W

X

MIX
Papier aus verantwortungsvollen Quellen
Paper from responsible sources
FSC® C105338

If you have any concerns about our products,
you can contact us on
ProductSafety@springernature.com

In case Publisher is established outside the EU,
the EU authorized representative is:
Springer Nature Customer Service Center GmbH
Europaplatz 3, 69115 Heidelberg, Germany

Printed by Libri Plureos GmbH
in Hamburg, Germany